Information Society and Development

Information Society and Development

The Kerala Experience

Edited by

Antony Palackal

Wesley Shrum

RAWAT PUBLICATIONS

Jaipur • New Delhi • Bangalore • Mumbai • Hyderabad • Guwahati

ISBN 81-316-0152-8

Published by
Prem Rawat for **Rawat Publications**
Satyam Apts, Sector 3, Jawahar Nagar, Jaipur 302 004 (India)
Phone: 0141 265 1748 / 7006 Fax: 0141 265 1748
E-mail: info@rawatbooks.com
Website: www.rawatbooks.com

New Delhi Office
4858/24, Ansari Road, Daryaganj, New Delhi 110 002
Phone: 011 2326 3290

Also at Bangalore, Mumbai, Hyderabad and Guwahati

Typeset by Rawat Computers, Jaipur
Printed at Chaman Enterprises, New Delhi

Dedicated to

LOYOLA COLLEGE OF SOCIAL SCIENCES
Trivandrum, Kerala (India)

Contents

Acknowledgements 9
Abbreviations 11
Glossary 14
Contributors 15

Introduction 21
Antony Palackal and Wesley Shrum

Section I
Circumventing the Digital Divide

1 Diffusion and Development: Kerala Experience of ICT for the Masses 41
Aruna Sundararajan

2 Akshaya: An Experiment in 'ICT For All' 52
T.K. Manzoor

3 ICT-Enabled Information Systems in Developing Countries: A Case Study 63
K.P. Sadasivan

4 Free Software in Kerala 77
V. Sasi Kumar

5 IT@School and Free Software in Education: The Kerala Model 96
Biju Prabhakar and Arun M.

Section II
Development Discourse

6 ICTs for Governance and Development: A Critical Review of Kerala's Recent Experiences 109
T. T. Sreekumar

7 Understanding Concepts of 'Development' and Linkages to Deployment Strategies for ICT for Development 133
Renee Kuriyan and Jennifer Bussell

8 The Globalization of Science: Is the Internet Changing the Careers of Female Researchers in Kerala? 151
B. Paige Miller

9 Who Has the Internet Empowered? Rethinking the Relationship between Women and ICTs in the Developing World 173
Meredith Anderson

10 Knowledge Production and Developmental Structures: E-science in a Developing World 200
Antony Palackal and Wesley Shrum

11 Information Technology for Kerala's Development 223
Jayan Jose Thomas

Index 230

Acknowledgements

As the dedication to our volume indicates, the students and faculty of Loyola College of Social Sciences (Trivandrum, Kerala) have been instrumental in the chain of events that led to this volume. Our indebtedness begins with the late and much-loved Father Dominic George, through Rev. M.K. George, the Principal, Rev. Pious Vachaparambil, the Rector and Manager, Rev. Joye James, the Provincial, Kerala Jesuit Province and Elizabeth Mathew, the HOD, R. Sooryamoorthy, Jose Boban, Saji P. Jacob, Lekshmi V. Nair the faculty members of the Department of Sociology, and to all the teaching faculty of the Departments of Social Work and Personnel Management.

Our colleagues in the World Science Project – even those who have not worked specifically in Kerala – are friends and collaborators who have enriched us both intellectually and personally, including Paul Mbatia in Kenya and Dan-Bright Dzorgbo in Ghana. R. Sooryamoorthy, before his departure for South Africa in late 2002, was instrumental in conducting the second wave of the Kerala study in 2000 as well as a study on mobile telephony. While the Loyola students who have given so generously of their time in surveying, coding, and interviewing are too numerous to mention individually, we would like to thank their North American colleagues Paige Miller, Meredith Anderson, Rick Duque, and Marcus Ynalvez.

The event summarized in this volume begin as one of four official side meetings that occurred in conjunction with the second phase of the World Summit on the Information Society that occurred in Tunisia in November of 2005. Along the way to that event, many played a role, including Bok Marais, Wiebe Bijker, Steve Coffee, Wayne Johnson, Barbara Waugh, Sheila Jasanoff, Paul Uhlir, Kathleen Cass, Carol Priestley, Peter Ballantyne, Klaus Brunnstein, Chen Huai, Keith Benson, Khalid Fourati, and Carthage Smith.

Needless to say, the book is the fruition of a collective endeavour and we are especially grateful to the authors for their willingness to contribute articles. We thank Rawat Publications for bringing out this volume. We must give first place, of course, to own loved ones – Susan, Ansu and Sachu.

Antony Palackal
Wesley Shrum

Abbreviations

APDIP	Asia-Pacific Development Information Programme
BOOT	Build-Own-Operate-Transfer
BPO	Business Process Outsourcing
BSNL	Bharat Sanchar Nigam Limited
B2C	Business to Citizen
BTM	Build-Transfer-Maintain
CAS	Chemical Abstracts Service C-DIT
CCOD	Current Contents On Diskettes
CCN	Computer-Communication-Network
CDAC	Centre for Development of Advanced Computing
CDIT	The Centre for Development of Imaging Technologies
CEO	Chief Executive Officer
CESS	Centre for Earth Science Studies
CLASS	Computer Learning And Studies in Schools
CMC	Computer Mediated Communication
CSIR	Council of Scientific and Industrial Research
CSO	Civil Society Organization
CTCRI	Central Tuber Crops Research Institute
CWLAN	Campus Wide Local Area Network
DfID	Department for International Development
DOE	Department of Electronics
DoT	Department of Telecommunications
DPI	Director of Public Instructions
DTH	Direct-To-Home

ERNET	Education and Research Network
FRIENDS	Fast, Reliable, Instant, Efficient, Network for Disbursement of Services
FLOSS	Free, Libre and Open Source Software
FSF	Free Software Foundation
GPSS	Gateway Packet Switching System
G2C	Government to Citizen
IAS	Indian Administrative Service
ICDS	Integrated Child Development Services
ICT	Information and Communication Technology
IIITMK	Indian Institute of Information Technology and Management, Kerala
IIT	Indian Institute of Technology
IKM	Information Kerala Mission
IMG	Institute of Management in Government
INSDO	Indian National Scientific Documentation Centre
ISRO	Indian Space Research Organization
IT	Information Technology
ITU	International Telecommunication Union
JCR	Journal Citation Report
KAU	Kerala Agricultural University
KBIP	Kerala Bureau for Industrial Promotion
KSSP	Kerala Sasthra Sahitya Parishad
KSTA	Kerala School Teachers Association
LIS	Library and Information System
LSP	Local Service Providers
LTSP	Linux Terminal Server Project
LUG	Linux User Group
MDGs	Millennium Development Goals
MGP	Modernizing Government Programme
MLA	Member of Legislative Assembly
Milma	Kerala State Milk Marketing Federation
MIS	Management Information System
MP	Member of Parliament
NGO	Non Governmental Organisation
NIB	National Internet Backbone
NIC	National Informatics Centre

NCSI	National Centre for Science Information
NCST	National Centre for Software Technology
PAD	Packet Assembler Dis-assembler
PPMS	Programme Performance Monitoring System
PPP	Public-Private Participation
PTA	Parent Teachers Association
PWD	Public Works Department
RENNIC	Research and Education Network of the National Informatics Centre
ROT	Receive Only Terminal
RRLT	Regional Research Laboratory, Trivandrum
SAC	Citizen Assistance Service Centers
SC	Scheduled Caste
SCERT	State Council for Educational Research and Training
SIET	State Institute of Educational Technology
SIT	Satellite Interactive Terminals
SIRNET	Scientific and Industrial Research Network
SIFFS	South Indian Federation of Fishermen's Societies
SPACE	Society for Promotion of Alternative Computing and Employment
SRO	Sub Registrar Office
SSLC	State School Living Certificate
S2S2	Swatantra Software Solutions and Services
ST	Scheduled Tribe
STD	Subscriber Trunk Dialling
STED	Science and Technology Entrepreneurship Development
TeNeT	Telecommunications and Computer Networking
UK	University of Kerala
UNDP	United Nations Development Programme
VICs	Village Information Centres
ViCTERS	Virtual Class Technology on Edusat for Rural Schools
VMEs	Virtual Micro Enterprises
VSAT	Very Small Aperture Terminal
VSNL	Videsh Sanchar Nigam Limited
VSSC	Vikram Sarabhai Space Centre
WLL	Wireless in Local Loop
WSIS	World Summit on the Information Society

Glossary

Akshaya	—	Unending
Alakal	—	Waves
Amma Ariyaan	—	To my Mother
Bhoomi	—	Land
e-bandham	—	e-contact
e-kendras	—	e-centres
Kudumbasree	—	Prosperity of the Family
Nadan pattu	—	Folk songs
Panchayat	—	Village Council
Sevanamudra	—	Sign of Service
Sutharya Keralam	—	Transparent Kerala
Swadeshi	—	Indigenous
Swathantra	—	Free
Vigyan	—	Knowledge

Contributors

Antony Palackal teaches in the Post Graduate Department of Sociology in Loyola College of Social Sciences, Kerala, India. He is the India Coordinator of the World Science Project that studies the impact of Internet on research communication in developing countries. Palackal obtained PhD from the University of Kerala for a doctoral dissertation in the area of globalization and culture. His latest books include *Grassroot Development: Establishing Successful Microenterprises–Kerala Experience* (with Murali Nair and Ralph Brody) and its Malayalam version. He has published many research and popular articles. He serves as a consultant and trainer to several NGOs and socio-political movements in Kerala. He offers classes, courses and seminars on topics of socio-cultural significance. His areas of specialisation include: Sociology of Science and Development, Social Movements, Cultural dynamics and Gender and Society. antonypalackal@yahoo.com

Wesley Shrum has been Professor of Sociology at Louisiana State University, USA since 1982 and Secretary of the Society for Social Studies of Science since 1987. His specialty is the sociology of science and technology. Dr. Shrum began the World Science Project in 1994 with a study of scientific networks in Ghana, Kenya, and Kerala. He has returned to Kerala regularly since that time and developed a love for its people and culture. With the spread of the Internet through these areas, he shifted focus in the late 1990s to the globalisation of science. (Manuscripts may be found at http://worldsci.net/global) His latest contribution is in the area of video ethnography and the preparation of academic movies. shrum@lsu.edu

Arun M. is a Free Software activist who was one of the founding members of Free Software Foundation of India, the third chapter of the International Free Software Foundation. He initiated the Malayalam GNU/Linux localization project, an attempt in developing Free Software for local languages and culture. Arun is involved in the IT@School Project that aims at bringing IT-enabled education to schools in Kerala. He has conducted studies on various e-Governance projects in Kerala and other ICT initiatives. He writes in local media on ICT and issues related to knowledge society. Currently he is working as the coordinator of the SPACE-Kerala initiative, a Government of Kerala sponsored initiative to promote Free Software and develop a new culture of IT enterprise in the State. arun@gnu.org.in

Aruna Sundararajan is a senior civil servant belonging to the Kerala cadre of the Indian Administrative Service. She has held key administrative posts both in the Government of India and the Government of Kerala. She is currently CEO of the CSC Project (Community Service Centres) of the Dept. of IT, Government of India. The CSC Scheme aims to create a national rural digital backbone and extend the benefits of ICTs to rural areas. The CSC scheme, which is one of the largest Public-Private Partnership (PPP) programmes in the world, envisages establishment of 1,00,000 rural community kiosks covering 6,00,000 of India's villages over the next 2 years. The project marks a key cornerstone of India's national vision to extend the benefits of IT to rural masses, thereby integrating them into the economic and social mainstream. Sundararajan, along with her team in ILFS, is spearheading the project implementation as the National Level Service Agency. arunaseg@rediffmail.com

Biju Prabhakar graduated in Engineering and Law from the University of Kerala and he is also a postgraduate in Business Administration from the same university. He worked as the Executive Director of IT@School Project, one of the largest IT projects of the Government of Kerala, in its inception phase and successfully implemented it. He also supervised the setting up of the ViCTERS, India's first broadband network on Edusat for schools in the State. He has been instrumental in revamping the existing Kerala government web site (kerala.gov.in), changing its architecture and design. Currently, he is the Deputy Collector of Revenue Department of Government of Kerala. He is also pursuing PhD in 'Project Management' in the University of Kerala. bijuprabhakar@gmail.com

B. Paige Miller is a sociology graduate student in her third year at Louisiana State University in Baton Rouge, Louisiana, USA. She received her master's degree in 2005 and is currently working towards her PhD. She graduated in sociology from Black Hills State University in Spearfish, South Dakota, USA in 2003. Since attending LSU she has worked on the World Science Project, an international collaboration that focuses on the impact of technology on the practice of science in developing areas. She specifically examines gender differences in the scientific careers of male and female researchers in Kerala, India. bmill@lsu.edu

Jayan Jose Thomas is a Research Fellow at the Institute of South Asian Studies, National University of Singapore (NUS). He has completed Ph.D. in development studies from the Indira Gandhi Institute of Development Research (IGIDR), Mumbai in 2003, and a Bachelor's degree in industrial engineering from Kerala University in 1995. Jayan's research so far has dealt with various aspects of Indian industrialization, issues of labour, technology and development, and the digital divide. While India will continue to be the focus of his future work, Jayan is keen to expand his research interests to other regions, particularly China. He hopes to contribute to the building of an interface between economics and other social sciences. His recent research papers have appeared in reputed journals including *World Development* and in edited volumes published by Oxford University Press and Palgrave Macmillan Ltd. He was awarded the 'Young Labour Economist Award' of the Indian Society of Labour Economics in 2003. Currently, Jayan is researching on three topics: globalization of innovation and pharmaceuticals and biotechnology industries in India and China; development of India's eastern region through South and East Asia collaboration; and, labour, capital and the rise of India and China. jayanjthomas@yahoo.co.in

Jennifer Bussell is a PhD candidate in Political Science at the University of California, Berkeley. Her dissertation investigates how political incentives affect information technology initiatives in developing countries. Prior to Berkeley she was a management consultant, advising major corporations in the United States, United Kingdom, and South Africa on key business issues, with an emphasis on the implications of digital technologies for business strategy. She earned her undergraduate degree in Anthropology at the University of Chicago. jbussell@berkeley.edu

K.P. Sadasivan has 34 years of professional experience as Head of Library/Technical Information Centre/Informatics of the R&D Labs of Defence Research & Development Organization (DRDO) and Council of Scientific & Industrial Research (CSIR), India. He is presently, Consultant for ICT Development in the National Institute for Interdisciplinary Science & Technology (CSIR), Trivandrum and Counsellor for the PG Programme (MLISc) of Indira Gandhi National Open University (IGNOU), Trivandrum Centre. He has delivered several invited lectures and seminars on various themes related to Information systems for Science and Technology and organized several national and regional conferences and seminars in Information Science and S&T Library Systems. Some of his significant achievements include: Introducing for the first time Direct Online Search Service of International Databases (DIALOG and STN) in Kerala at RRL Trivandrum, E-Mail system facility (SIRNETM) and SciFinder Service (CAS, Ohio). He played a significant role in introducing the CSIR e-Journals Consortia Initiative, the first major EJC Programme in the corporate sector in India. He had been the Nodal Officer for several projects like CSIRNET, Y2K Task Force for CSIR at RRLT, CSIR e-Journals Consortia Project, CSIR ICT Development Project, etc. kps31147@yahoo.com

Meredith Anderson is a doctoral candidate in the Department of Sociology at Louisiana State University in Baton Rouge. She is a qualitative methodologist whose research interests lie in the intersection of development, science and technology, and gender. Meredith has conducted research in Kerala since 2003 and is currently beginning work on her dissertation, which will assess the impact and sustainability of Kudumbashree, a poverty eradication initiative targeting poor women in Kerala. smee_me@yahoo.com

Renee Kuriyan is a PhD Candidate at the University of California, Berkeley focusing her dissertation research on the political economy of information and communication technologies for development. Her undergraduate degree is in Anthropology from Middlebury College and she has a Masters degree in Public Affairs from Princeton University's Woodrow Wilson School with a focus on international development. She has extensive experience conducting ethnographic research on a range of technology and development topics in Africa (Kenya and Tanzania) and India. This includes projects with the World Bank, UNDP, UNEP, multinational corporations and NGOs. rkuriyan@berkeley.edu

T.K. Manzoor is Director of Akshaya, an ICT dissemination project promoted under the State Government of Kerala. The project has created a unique brand of public-private participation in establishing 3000+ broadband enabled e-centres known as Akshaya e-centres spread over even the remotest part in the State, providing a wide range of services and making way to many future opportunities. He joined Akshaya in 2005 and heads the activities of the project at the State level. Prior to his current position, he was working as Director of Science & Technology Entrepreneurship Development (STED) Project during 2002-2005. STED Project was initiated by the National Science & Technology Entrepreneurship Development Board under the Department of Science & Technology of the Government of India with objectives of entrepreneurship development and employment generation through Science & Technology inputs in the State of Kerala. He also played a crucial role in the conceptualization of Akshaya Project, while working in STED Project in 2002. He is a Bachelor in Technology (B.Tech) with Masters degree in Business Administration. He has experience in the areas of entrepreneurship development, business/project appraisal, ICT Development etc. in India and Sultanate of Oman. manzoortk@yahoo.com

T. T. Sreekumar is Assistant Professor at the National University of Singapore, Singapore. He has published articles and reviews in major international journals and his papers have also appeared in several international conference proceedings. Recent publications include "ICTs for Rural Poor: Civil Society and Cyber-libertarian Developmentalism in India", pp 61-87 in Parayil. G (ed.) *Political Economy and Information Capitalism in India: Digital Divide,*

Development Divide and Equity, UK: Palgrave Macmillan, 2006; "Interrogating Development: New Social Movements, Democracy and Indigenous People's Struggles in Kerala", pp 215-57 in Tharamanaglam J. (ed) Kerala: *Paradoxes of Development and Public Action*. New Delhi: Orient Longman (with G. Parayil) and "Combating Child Labour in India 's Services Sector", *South Asian Journal,* No. 14 [October-December] (with V Gayathri). cnmsttp@nus.edu.sg

V. Sasi Kumar obtained M.Tech in Physical Engineering from the Indian Institute of Science. He joined the Atmospheric Sciences Division of the Centre for Earth Science Studies (CESS), Thiruvananthapuram in 1979. He later obtained a PhD in Physics from the Cochin University of Science and Technology for his work in atmospheric electricity. His current research areas include rainfall studies and cloud physics. He is actively involved in promoting Free Software and is one of the directors of the Free Software Foundation of India and also of SPACE. He has been involved in the IT education programme at school level in the State and is currently involved in the process of curriculum revision and introducing IT Enabled Education at the high school level. He has written a number of popular science articles in English and Malayalam and has also contributed to the publication of a popular science magazine called *Science India*. vsasi@hotpop.com

Introduction

ANTONY PALACKAL AND WESLEY SHRUM

The digital revolution in the fields of information and communication, a key feature of globalization dynamics, has extended the frontiers of the global village with a profound impact on how the world functions and interacts. Over the decades, ICT has triggered shifts affecting education, culture, society and many other spheres of our lives. Access to information and knowledge facilitated by ICT is increasingly impacting patterns of learning, cultural expression and social participation, as well as providing opportunities for development (Kleiwachter and Stauffacher 2005; Khan Waheed Abdul 2005). These developments have forged new ways to create knowledge, educate people and disseminate information. They have restructured the way we conduct economic and business practices, run governments and engage politically. They have provided for the speedy delivery of humanitarian aid and healthcare, and a new vision for environmental protection. It is now accepted that ICTs have enormous potential to improve people's economic, social and cultural well being in a knowledge-based, digital economy. The information and communication revolution penetrates all aspects of our lives.[1]

It has been claimed that ICTs have a major role to play, especially in the less developed regions of the world, as they help connect individuals, small companies or groups or farmers and artisans in the poor and isolated areas of the world, integrating them into national or even global markets (Anabel Quan-Haase and Barry Wellman 2002). ICTs foster new forms of social organization and networks that are less bounded than traditional bureaucratic hierarchies. Computer Mediated Communication (CMC) aids rapid communication and information access making easier, inexpensive and convenient communication with far-flung communities. Coupled with a low operating cost and the ability to communicate while the other person is not immediately available, CMC can create an enhanced ability to maintain spatially dispersed, sparsely knit, and interest-based relationships. Hence it is argued that CMC has facilitated a mutation in the form of community because CMC facilitates collaboration that is both hyper-connected and glocal (Ibid: 281). Hence, computer networks are viewed as inherently social networks, linking people, organizations and knowledge. The proliferation of computer networks has facilitated a de-emphasis on group solidarities at work and in the community, affording a turn to networked societies that are loosely bounded and sparsely knit (Quan-Haase Anabel and Barry Wellman 2002).

The advent of ICT as a social phenomenon has been viewed through many theoretical lenses. In the backdrop of the dramatic increase all over the world in the use of Internet for communication and information retrieval since the 1990s,[2] some studies have employed the concept of 'social capital' as a heuristic tool in examining the vitality of a locale in terms of its participation in social networking and development (Putnam 1993, 1996, 2000; Quan-Haase and Barry Wellman 2002). There are two complementary usages of the 'social capital' concept: First, it is used to denote social contacts (interpersonal and organizational communication patterns and networks). Second, it is used to mean civic engagement, the degree to which people become involved in their community in all vistas of social life, both actively and passively.

Studies on the relationship between the Internet and social capital reveal varied and even contrasting effects (Quan-Haase and Wellman 2002):

1. The Internet may increase social capital by providing the means for inexpensive and convenient communication with far-flung communities of shared interest. Barlow et al. (1995) argues with the Internet's low costs and often-asynchronous nature, it leads

to a major transformation in social contact and civic involvement away from local and group-based solidarities and towards more spatially-dispersed and sparsely-knit interest-based social networks.

2. The Internet may diminish social capital through its entertainment and information capabilities, as it draws people away from family and friends. Further, by facilitating global communication and involvement, it may reduce interest in the local community and its politics (Nie et al. 2002; Nie 2001).
3. The Internet supplements social capital by blending into the lifeworld. In this view, Internet is regarded as another means of communication to facilitate existing social relationships and follow patterns of civic engagement and socialization. People use the Internet to maintain existing social contacts, adding electronic contact to telephone and face-to-face contact.

Our approach in this volume is not to portray ICTs exclusively as instruments for enhancing networks or transforming social capital, but to include them as tools for social development. Hence, as the title of the volume suggests, we value the potential given official recognition at the World Summit on the Information Society[3] (WSIS). In the late nineties, the General Assembly of the United Nations considered the economic and social effects of globalization and the policies needed to address them. They recognized that the information revolution was having a profound impact on the global economy and development. The Assembly agreed that information and communication technologies should be put at the service of development (Khan Sarbuland 2005: 41). The first phase of the World Summit on the Information Society, which met in Geneva in December 2003, gave shape to a shared vision and put forward a concrete plan of action to build a development-oriented, equitable and inclusive Information Society "where everyone can create, access, utilize and share information and knowledge, enabling individuals, communities and peoples to achieve their full potential in promoting sustainable development and improving their quality of life" (WSIS Declaration of Principles No.1). The idea was that since ICTs are knowledge-intensive, using them to transfer knowledge and market-driven skills would be a powerful instrument for fighting poverty and deprivation. The ICT revolution was hailed as the enabling force for developing countries to become more active participants in the global development. As the Regional

Human Development Report of UNDP observed: "Developing countries the world over have made the widespread use of ICTs a central feature of their development agenda" (1).

Notwithstanding these well-intended deliberations by international forums, the reality is that today's information and communication society is characterized by imbalances and inequalities, popularly known as the 'digital divide'. The digital divide involves the gap between individuals and societies that have the resources to participate in the information era and those that do not, a gap that leads to further differences in the newly critical resources of ICT hardware, connectivity and content. In the context of the unprecedented developments in information and communication technology, the world today seems to witness two opposite trends. On the one hand explosive changes in science and technology offer uncommon opportunities for health, food, water, work, energy, and literacy and above all communication and information to many. On the other, much of humankind is living under conditions of poverty, hunger, and deprivation, feeling a sense of social exclusion and injustice.

At the epicentre of this alarm is the expanding chasm between those with and without access to the revolutions in the field of ICTs. There is a growing concern about the world's technological haves and have-nots. What is more worrisome about this gap is that it distances the poor and marginalized people of the world by greater degrees than ever before from its advantaged citizens. The revolutionary developments in ICT appear to have brought new contradictions such that the digital divide is actually several gaps in one:

1. A technological divide – great gaps in infrastructure and its access;
2. A content divide – web-based information irrelevant to the real needs of people;
3. A gender divide – women and girls have less know-how, use and access to ICT than men and boys (although this now appears to have closed in some developed areas);
4. A commercial divide – e-commerce is linking some countries and companies ever more closely, while others run the risk of marginalization.

There are other obvious social, economic and other disparities and obstacles that affect a country's ability to take advantage of digital opportunities. Hence, the digital divide reflects the broader context of international social and economic relations. Kofi Annan, the former UN Secretary General has said, "the ICT age has dawned, but not yet for all. Let us show that we can unite the great promise of ICT with the needs of the poor".[4]

Affirming the importance of inclusiveness to knowledge societies, the WSIS Declaration of Principles stated unambiguously that " ... we are fully committed to turning this digital divide into a digital opportunity for all, particularly for those who risk being left behind and being further marginalized" (No. 10). All persons, without distinction, must have access to its benefits and must be empowered to create, receive, share and utilize information and knowledge freely (Abdul Waheed Khan, 2005: 19). Yoshio Utsumi, the Secretary General, World Summit on the Information Society and the Secretary General, International Telecommunication Union (ITU) added: "bridging the digital and knowledge divide and avoiding the inequities of the past has taken on unprecedented urgency for the Information Society as it moves from the past into the future".[5] The second phase of the World Summit on the Information Society in Tunis in November 2005, against the backdrop of which this edited volume has been originally conceived, therefore seemingly made a serious attempt to grapple with the contradictions and challenges of the information-technology-driven times, besides demonstrating the progress made in technology development since the 2003 Geneva Summit. This edited volume is an endeavour to portray the attempts at circumventing the digital divide in the State of Kerala, a distinctive developing region in the southwestern part of India.

The volume was conceived at the conclusion of the Kerala Special Session of the international conference in Tunisia that occurred in association with Phase II of the World Summit on the Information Society (Shrum et al. 2007). 'Past, Present and Future of Research in the Information Society' was held on the Mediterranean coast near Tunis in November 2005, registered as an official side event of the WSIS on science, engineering and technology (Shrum 2007; http: //worldsci.net). Kerala was indeed the only political or geographical area to receive a session of its own in the entire conference, as a result of the unique status that the State of Kerala possesses within the development community (Palackal et al. 2007).

Kerala holds a distinctive position in the developmental map of the world. With a current population of 31,838,619 people living in 38,863 square kilometres, Kerala continues to be the focus of attention for both Indian and international scholarship as a region in the developing world with a unique pattern of development. The 'Kerala Model', as it is popularly known, has been a much-debated topic in the academic circles, for it appears paradoxical, exhibiting a developmental trajectory with high social achievements[6] on a weak economic base (Ramachandran 1996; Oommen 1992, 1999; Sen 1997; Kannan 2000; Shrum and Ramanathaiyer 2000). The State, with its unique socio-economic and demographic characteristics, is comparable in many respects with developed nations. The Nobel Laureate Amartya Sen wrote: "Kerala, despite its low-income level, has achieved more than even some of the most admired high-growth economies such as South Korea (Sen 1997: 7). On the other hand, notwithstanding such enviable progress and unique features, Kerala remains to be one of the poorer regions among the states in India. Yet the bleak economic scenario of the State contradicts its achievements. Many studies later brought to surface the fact that the Kerala Model faces several problems and shortcomings that taken together can be called a crisis[7] (Kannan 1986; George 1993; Mohan 1994; Franke and Chasin 2000). With a per capita income below the national average, Kerala exhibits poor economic growth but a high standard of living. In sharp contradistinction to its weak economy, a high level of consumption marks a living standard for its people. In recent years, Malayalis engaged in forms of consumption that resemble the cosmopolitan cultural forms in other parts of the contemporary world. Amidst this ironic developmental trajectory, Kerala recently experimented with the world's first sub-national campaign, known as the People's Campaign for Decentralized Planning. It became a model of participatory democracy and participatory development, which is christened as the New Kerala Model (Isaac and Franke 2000). Defined in the broad sense of the campaign, it was an attempt to conceptualise, operationalize and institutionalize a multi-level, people-centred planning process suitable to the regional specificities of Kerala. The primary objective of the Campaign has been to motivate and empower local self-governments to take up new challenges of development planning (Nair et al. 2007).

With the backdrop of these paradoxical, but high priority developmental concerns, the State sought front stage among the ICT initiatives in India. The State's IT policy (2001) acknowledged in its preamble that it "endeavours to delineate a strategy for harnessing the opportunities and the resources offered by IT for the comprehensive social and economic development of the State". It also states that "the growth of Kerala in coming years will be increasingly driven by the knowledge and service-based sectors, where ease of information transactions will be a key determinant of success". The mass base enjoyed by the media and the penetration of communication technologies are definite advantages in making the State of Kerala a 100 per cent *'internetised'* State – truly, an Information Society. It appears that the Government of Kerala understands the enormous potential of ICTs not only as a tool for improving governance and creating more jobs, but more significantly, as a means to greatly enhance the standard of living of the people. The latest draft of the Information Technology policy (2007) proposes to make Kerala a cradle of knowledge-workers.[8]

The present volume explores the nature and effects of information and communication technologies with particular emphasis on development programmes and the scientific research and engineering communities. Since 1994 the World Science Project (http: //worldsci.net) led by an international team of scholars and primarily responsible for the "Past, Present and Future" meeting in Tunisia, has been involved in a longitudinal study in the State of Kerala. We open up a line of enquiry into the ways in which the introduction of new information and communication technologies (ICTs) have changed communication patterns and relationships and the process of knowledge production by those engaged in research activities. Since the advent of the Internet, there are many who have claimed that the production of knowledge has become a global enterprise, with unlimited possibilities for communication, collaboration and data transmission across international boundaries. Three articles in the present volume (chapters 8, 9 and 10) contain theoretical and empirical studies from this long term project, shedding particular light on the gender and digital divide within the research community of Kerala. Our sociological gaze fixes on the structure and ramifications of ICT intervention in terms of communication, collaboration and social development of a developing region in both the public sector and the scientific enterprise. The overall objective of the endeavour is to document the e-Science initiatives in Kerala vis-à-vis development within the larger framework of the ongoing discourse on global structural changes in development and knowledge production.

The specific tasks of this volume include the following:

1. Reviewing the various e-Science programmes and practices of the State, including the diffusion of Free Software from a historical and policy perspective in the context of the distinctive developmental scenario of the State,
2. Appraising the Akshaya programme in Kerala, an initiative of the government of Kerala that aims at making Kerala an IT literate State,
3. Examining the different philosophies of development and project goals of ICT For Development initiatives and problematizing the issue of 'development',
4. Investigating diachronically the nature and structure of knowledge production in Kerala vis-à-vis ICTs,
5. Exploring the level of diffusion of ICTs and its effect on the resource acquisition capabilities and career prospects of female scientists in Kerala.

As such, we have not attempted a comprehensive overview of ICT activities within Kerala, but focused our selection of authors and essays on those that were thematized at the World Summit, and particularly the Kerala Special Session of "Past, Present and Future of Research in the Information Society" (Palackal et al. 2007; Shrum 2007). We begin with Aruna Sundararajan, former Information Technology Secretary of Kerala, who provides an introduction to the major ICT initiatives in Kerala, several of which have received international recognition. Noting that no one doubts the potential of ICTs to promote development, but poor infrastructure, inequitable access, the lack of basic skills and relevant content, as well as the obvious scarcity of resources are barriers so deep-rooted in the Indian context as to seem insurmountable. She attacks this issue through detailed introductions to three initiatives, two of which deal directly with governance issues while the third is directed towards empowerment of the poor. FRIENDS is a relatively simple access point with the goal of improving State service levels. Sharply focused on the interface between citizens and government, the programme has logged successes in reducing access time for certain services. The Information Kerala Mission is the largest and most comprehensive local computerisation project in India. Its goal is to transform administration at the grassroots through networking of 1215 local self-government institutions. FRIENDS and IKM, extensive though they are, promote a single portfolio of services. Akshaya is the most widely heralded ICT dissemination initiative in the State and one of the most ambitious anywhere, building on Kerala's earlier 'universal literacy' programme. In the broadest sense, Akshaya seeks to empower an entire community across the entire State, including backward and remote segments, with the promise of disseminating ICT tools to each household for use in ways that would be relevant to their individual developmental issues. Sundararajan argues that the manner in which the Akshaya Project develops, its failures and successes, will provide important lessons for understanding the ability of ICTs to transform ordinary lives. Her introduction is critical to the more detailed presentation and analysis of Akshaya in chapters 2, 7, and 8.

T.K. Manzoor serves as the Chief Executive Officer, Akshaya. His commitment and enthusiasm for the award-winning Akshaya Project of the Kerala State Government are reflected throughout his essay, which serves as a detailed description of the project in its first phase. Given the attention showered on the project, the editors regard Akshaya as a focal point of Kerala IT in the State sector in partnership with the private sector. Manzoor guides us through the multifarious goals of the project from its inception through the establishment of centres in the Malappuram district of Kerala. This area sends a large number of guest workers abroad and could be significantly helped by the information and communication capacities that e-centres might provide. Akshaya seeks to establish access points throughout the State, with one centre for each 2500-3000 households, to train one person from each household in computer skills, and to provide local language content of relevance to improving the daily lives of people. Manzoor discusses the role of local governments and promotional activities that are used in beginning the rollout of the centres, provides a socio-demographic overview of the entrepreneurs that play a key role in their local management, and highlights the important role of women, who constituted one third of the original group of entrepreneurs. While Manzoor notes that the success of the project has been a motivation for the current expansion throughout all of Kerala, he does not provide empirical evidence for this success, beyond a discussion of the awards the project has received.

K.P. Sadasivan's essay excels in specificity, more than any in this volume, as a personal and historical account of the evolution of ICTs in a national research laboratory based in Trivandrum. One of the main promoters himself, Sadasivan provides an insider account of communications developments in the Regional Research Laboratory where for three decades he directed the Library and Informatics Division. He lends insight into the management approach of the lab and the proactive behaviour of his own division in the process of adoption by research scientists of novel information and communication systems. The financial issues and national policy decisions that led gradually to relatively sophisticated infrastructural facilities: bibliographic services in the 1980s, networked personal computers in the 1990s, and finally the web-enabled information resources in the past decade. Sadasivan provides two detailed case studies through his personal accounts of (1) resistance to services sought by the Library and Informatics Division, which took the form of active attempts to block email services, and (2) the successful introduction of online search services. Finally, he presents some evidence on productivity changes and stories to support his feeling that the provision of new information technologies has improved the performance efficiency of the Lab.

We turn next to two articles focusing on Free Software, as the movement seems to have found an exceptional home in Kerala. V. Sasi Kumar begins with an examination of the meaning of the "four freedoms" of Free Software and the history of these developments in Kerala. He describes the introduction of the TEX programme at the University of Kerala and the establishment of River Valley Technologies in 1995. He recounts the role of C.V. Radhakrishnan, Satish Babu, and M. Arun as well as the gradual emergence of interest in GNU/Linux. Well-publicized conferences and visits by Richard Stallman were important to this interest, and the founding of the Free Software Foundation of India and an association of Free Software developers. Sasi Kumar describes the first successful campaigns for Free Software in the public works and education departments of the State, and the reasons for their success, offering implicit guidance for those who would want to conduct such campaigns. In a variety of applications, including computerized examinations, development projects, and business, he examines the perils and potentials of moving towards this non-proprietary mode of IT. He concludes with remarks on the unique nature of Kerala society and its new role as a home for Free Software.

Biju Prabhakar and Arun M. follow with a detailed description of one of the most exciting developments in Free Software in Kerala, the IT@School programme, which began in 2002. They view this as a major effort to circumvent the digital divide that will be a model for the rest of India. The use of Free Software in the schools is likely to reduce costs and improve learning, illustrating how Free Software can be used to the advantage of the people at large. IT@School seeks to bring computer training to high school students and is now one of the largest IT education programmes based in schools. Not only did it enhance traditional teaching, but induced changes in curriculum, content and pedagogy. Prabhakar and Arun believe that prior failures in the introduction of technology in the classroom from the 1960s to the 1990s have been overcome, at least partially, with the new approach. With IT@School, advocacy campaigns were combined with attempts to address resource constraints. However, one of the most important events was the objection by advocates of Free Software in Kerala to the use of proprietary software in the public education system. With the switch to a Free Software model, the authors believe that IT@School will return Kerala to the forefront of socially responsible education in India.

We follow this optimistic view of a Kerala IT programme with the more pessimistic account by T.T. Sreekumar. This article examines e-governance projects we have previously introduced, FRIENDS and Akshaya, concluding that they are, thus far, failures. Sreekumar notes that India was one of the earliest to introduce ICTs in development administration, beginning in the 1970s, before it became a subject of much international attention. He compares Kerala with Bihar, two states that differ widely in some respects but are similar in the progress of in e-governance. One of Sreekumar's most interesting findings is that while Bihar, with its department-centred approach, provides only a small role for civil society organizations in its e-Governance programme, even the socially focused programmes of Kerala are relatively antipathetic to such inclusions. Most of the analysis concerns FRIENDS, a programme that serves as an example of stagnation and negligence, owing to the apathy of the government departments that should lead in offering services. By comparison with a similar programme of Citizen Assistance Service Centers (SAC) in Brazil, Sreekumar comes to the view that the FRIENDS fails to match it in content and coverage.

Renee Kuriyan and Jennifer Bussell examine three specific cases of rural kiosk projects in India, focusing on the different definitions of 'development' they employ and the intended recipients of ICT access. Their first case study focuses on the Datamation Foundation where 'development' is defined largely in terms of social development characteristics with a specific attention to gender empowerment through the use of ICTs. Its special feature as a kiosk initiative is to emphasize characteristics and content to support women's development, with minimal services for men. However, the programme is dependent on subsidies for its existence and remains small. By contrast, the N-Logue kiosk programme places explicit emphasis on economic development in India's rural areas. Local entrepreneurs own and operate the kiosks and provide services they believe will make the business sustainable in their local market. They operate without subsidies, but there is very little attempt to target disadvantaged groups for access to the kiosks. Again, as in previous chapters, Akshaya is the focus in Kerala, but this time as a contrasting kiosk initiative that unites social and economic development goals. As in N-Logue, entrepreneurs are asked to initiate the centres, but the State subsidized e-literacy training in order to build a market: all citizens, in principle, should have the skills to make use of the kiosks. Kuriyan and Bussell conclude that in practice there are serious challenges for both the Kerala government and the kiosk entrepreneurs themselves to balance the two goals. In essence, there is a tradeoff between the commercial viability of these businesses and the provision of benefits for disadvantaged groups. They conclude that the type of 'development' evoked in a project affects its ultimate outcomes.

B. Paige Miller explores the changes that have occurred in female researchers' careers from 1994 to 2000 by addressing the impact of the Internet on the productivity levels and professional contacts of women in Kerala, India. While it is theorized that the Internet will globalize the practice of science by removing the constraints on women's physical movement enabling them to create contacts between themselves and other professionals, few empirical studies have explicitly examined this relationship. Using 393 respondents drawn from universities and national research institutes in Kerala, she analyzes the gender differences in women's travel experiences, access to the Internet and email, foreign and domestic contacts and productivity in foreign and national journals. Miller finds that access to personal computers and email increased, but the impact of these resources on women's careers has thus far been uneven. Over time and compared to men, they are more locally oriented, report fewer foreign ties, and their publication in foreign journals has decreased. At the same time, though, they report fairly stable domestic contacts and increased publications in national journals. Internet diffusion to Kerala, then, appears to be associated with increased activity in the local scientific arena but not greater integration into global science. While the data used to reach these conclusions are impressive in many respects, the second wave of the study was conducted in 2000, such that recent changes are not included.

Meredith Anderson asks whether information and communication technologies (ICT) are actually functioning to promote gender equality in the developing world? We are pleased to publish this paper, which is an early version of other work that has appeared in various forums, and was much debated at the Tunis World Summit event, "Past, Present, and Future of Research in the Information Society." Anderson rehearses the view taken by much of Western development discourse, which views the adoption of ICT by developing nations in an extremely optimistic manner. She reviews recent feminist critiques of western development policies as ethnocentric in nature and seldom consistent with local context. Anderson's empirical work focuses on the diffusion of ICT among women scientists in Kerala over the past decade. She concludes that the patrifocal social structure remains firmly in tact, but that Indian women scientists have taken advantage of professional opportunities available through the Internet to circumvent its limitations. By doing so they simultaneously advance their disciplines and promote social equality for women.

Our own article in this volume follows these two by Miller and Anderson focusing on Kerala women scientists. Miller compares women and men scientists from 1994 to 2000, while Anderson's qualitative data are slightly later. The consistency of our results with these two studies, particularly with respect to gender, reinforced our feeling that we should make them public in this early version. We hasten to emphasize that our results here are preliminary, since data cleaning and coding are still in progress as this volume goes to press. The dramatic increase in computer access and Internet connectivity reported by Miller continues, such that fewer than 10 per cent did not have full access. What is striking is that the proportion of women scientists with the doctoral degree is now higher than men, but women are still less likely to acquire an advanced degree from abroad and, if anything, there appear to be fewer women going abroad. Other results suggest gender remains a durable source of inequality. Among the few non-users, most were women, and organizational localism is still pronounced. Most important, the number of local publications by women has increased while the number of international publications has actually decreased since 2000, even as they have experienced greater Internet access. Surely, this is cause for concern.

Jayan Jose Thomas concludes this volume with a realistic assessment of Kerala's performance in the information technology industry to date and the conditions that would be required for progress. It seems, on the one hand, that given a large and educated workforce, Kerala is ideally positioned for rapid economic growth in a knowledge economy featuring ICTs. Thomas concludes, however, that the record so far is poor, and that the Kerala industrial environment has been characterized by technological backwardness and low diversification. With little assistance from the Central Government, a large investment by the State will be required. This might be in the form of a collaboration between government, industry and academic institutes in both information technology and biotechnology. One bright spot, noted by Thomas and others, is the diffusion of information technology in rural areas, a development facilitated by the rural-urban continuum characterizing the State.

Notes

1. Preface to *The World Summit on the Information Society: Moving from the Past into the Future, ICT Task Force Series 8*, (eds.) Daniel Stauffacher and Wolfgang Kleinsachter. New York: United Nations Information and Communication Technologies Task Force, 2005.
2. Although there are no reliable data on the size of the world's online population, estimates suggest that the number of Internet users

around the globe has surged from 4.4 million in 1991 to 10 million in 1993, to 40 million in 1995, to 117 million in 1997, to 277 million in 1999, to 502 million in 2001, and to more than 600 million in 2002. Thus, the global penetration rate of the projections for 2004 place the number of global Internet users between 700 million and 945 million; Chen and Wellman. 2004: 40.

3. WSIS Declaration of Principles No. 9 stated thus: "We are aware that ICTs should be regarded as tools and not as an end in themselves. Under favourable conditions, these technologies can be a powerful instrument, increasing productivity, generating economic growth, job creation and employability and improving the quality of life of all."
4. Cited by Khan Sarbuland, Executive Coordinator, United Nations ICT Task Force in his essay "WSIS and the United Nations ICT Task Force" in *The World Summit on the Information Society: Moving from the Past into the Future*, ICT Task Force Series 8, New York: United Nations ICT Task Force. 2005. p. 41
5. See 'Preface' to *The World Summit on the Information Society: Moving from the Past into the Future*, ICT Task Force Series 8, New York: United Nations ICT Task Force, 2005. p. xi
6. To mention a few unique features of Kerala: the State has the highest number of literates in the country among both men and women; the sex ratio is in favour of women, contrary to the common all India trend; Kerala has the lowest birth rate and death rate in the country; infant mortality rate too is the lowest which is comparable with that of Washington; fertility is one of the least rates recorded – less than two per couple; so is the maternal mortality rate: two for 1000 deliveries, in sharp contrast to forty-five in India; life expectancy at birth is well above the all India average; the average morbidity rate too is an all-time low.
7. Stagnation in agricultural and industrial production, widespread reduction in productivity levels, acute power shortages, out migration of industries and skilled labour, huge unemployment rate, all round reduction the quality of services and the debilitating fiscal crisis of the State etc. were identified as some facets of the alleged crisis. *The Economic and Political Weekly* devoted three issues to the theme 'Kerala Economy at the Crossroads' (*EPW* 25: 35-36 and 37, 1 and 8 and 15 September 1990). The Crisis of the Kerala Model was the major theme of the August 1994 First International Congress on Kerala Studies, held in Thiruvananthapuram. Hundreds of papers and discussions at the Congress reaffirmed suspicions and elaborated on the nature of the problems Kerala faces.

8. The major proposals include establishment of an International Centre for Free Software and Computing for Development and ITES Training Centre and extension of Internet to all educational institutions and villages by 2010 (*The Hindu*, January 18, 2007: 1). The Kerala State IT Mission is set to introduce its e-krishi project to 7 more districts following the instant success of the pilot project in Malappuram district. Farmers of those seven districts, where the Akshaya e-literacy project is being implemented after Malappuram will be able to sell their produce through Internet from next month (*The Hindu*, March 17, 2007. p. 11).

References

Barlow, J.P., Birkets, S., Kelly, K., and Slouka, M. 1995. "What are we doing on-line"? *Harper's*. August 291, 35-46

Chen, Wenhong and Wellman, Barry. 2004. "The Global Digital Divide – Within and Between Countries". *IT & Society*, 1 (7), 39-45

Franke, R.W. and B.H. Chasin. 2000. *Is the Kerala Model Sustainable? Lessons from the Past, Prospects for the Future*. London: Zed books.

George, K.K. 1993. "Limits to Kerala Model of Development: An Analysis of Fiscal Crisis and its Implications". Trivandrum: Centre for Development Studies.

Isaac, Thomas and R.W. Franke. 2000. *Local Democracy and Development – Peoples Campaign for Decentralized Planning in Kerala.* New Delhi: Left Word Books.

Kannan, K.P. 1986. "Kerala's Economy at the Crossroads?" *Economic and Political Weekly*, 25 (35-36), 1951-56.

—. 2000. "Poverty Alleviation as Advancing Basic Human Capabilities: Kerala's Achievements Compared". In Govindan Parayil (ed.) *Kerala: The Development Experience.* London: Zed Books.

Khan Sarbuland. 2005. "WSIS and the United Nations ICT Task Force". *The World Summit on the Information Society: Moving from the Past into the Future,* ICT Task Force Series 8, New York: United Nations ICT Task Force.

Khan, Waheed, Abdul. 2005. "WSIS and UNESCO". *The World Summit on the Information Society: Moving from the Past into the Future,* ICT Task Force Series 8, New York: United Nations ICT Task Force.

Kubicek, H. 2004. "Fighting a Moving Target". *IT & Society* 1 (6), 1-19

Mohan, R. and C. Panth. 1982. "Morphology of urbanisation in India – Some results from 1981 Census". *Economic and Political Weekly*. XVII (38-39).

Nair Murali, Ralph Brody and Antony Palackal. 2007. *Grassroot Development – Establishing Successful Microenterprises: Kerala Experience*. Cleveland: Human Networks Publication

Nie, N.H. 2001. "Sociability, Interpersonal Relations, and the Internet: Reconciling Conflicting Findings". *American Behavioural Scientist*, 45(3), 426-437

Nie, N.H., Hillygus D. S., and Erbring, L. 2002. "Internet use, interpersonal relations and sociability: A time diary study". In B. Wellman & C. Haythornthwaite (eds.). *Internet and Everyday Life*. Oxford: Blackwell, 215-243

Oommen, M.A. 1992. *The Kerala Economy*. New Delhi: Oxford & IBH Publishing Company.

—. 1999. *Kerala's Development Experience* (Vols. 1 & 11). New Delhi: Concept Publishing.

Palackal, Antony et al. 2007. "ICT and the Kerala Model". In Wesley Shrum, Keith Benson, Wiebe Bijker, and Klaus Brunnstein (eds.) *Past, Present and Future of Research in the Information Society*. New York: Springer, pp. 225-43

Putnam, R.D. 1993. *Making Democracy Work: Civic Traditions in Modern Italy*. Princeton, NJ: Princeton University Press.

—. 1996. "The strange disappearance of civic America". *The American Prospect*, 24, 34-48.

—. 2000. *Bowling Alone: The Collapse and Revival of American Community*. NY: Simon & Schuster.

Quan-Haase Anabel and Barry Wellman. 2002. "How does the Internet Affect Social Capital". In Marleen Huysman and Volker Wulf, (eds.). *IT and Social Capital*. University of Toronto

Quan-Haase, A., Wellman, B., Witte, J., and Hampton, K. 2002. "Capitalizing on the Internet: Social Contact, Civic Engagement, and Sense of community". In B. Wellman & C. Haythornthwaite (eds.). *Internet and Everyday Life*. Oxford: Blackwell. 291-324

Ramachandran, V.K. 1996. 'On Kerala's Development Achievements' In Jean Drèze, and Amartya Sen (eds.). *Indian Development: Selected Regional Perspectives*. New Delhi: Oxford University Press.

Sen, Amartya. 1997. "Economies and the Family." In Patricia Oberoi (ed.) *Family, Kinship and Marriage in India*. Delhi: Oxford University Press.

Shrum, Wesley and Ramanathaiyer, Iyer. 2000. "Knowledge, Democratization, and Sustainability: The Kerala Model of Scientific Capacity Building". In Govindan Parayil (ed). *Kerala: The Development Experience*. London: Zed Books. 151-177.

Shrum, Wesley. Keith, Benson., Wiebe, Bijker. and Klaus, Brunnstein. (eds.). 2007. *Past, Present and Future of Research in the Information Society*. New York: Springer.

Shrum, Wesley. 2007. "Introduction: Learning from the Past, Present and Future". In Wesley Shrum, Keith Benson, Wiebe Bijker, and Klaus Brunnstein (eds.). *Past, Present and Future of Research in the Information Society*. New York: Springer. 1-12.

Stauffacher, Daniel and Kleinsachter, Wolfgang. 2005. "Introduction". *The World Summit on the Information Society: Moving from the Past into the Future*, ICT Task Force Series 8, New York: United Nations ICT Task Force.

UNDP. 2005. *Promoting ICT for Human Development in Asia*. India.

I

Circumventing the Digital Divide

1

Diffusion and Development

Kerala Experience of ICT for the Masses

ARUNA SUNDARARAJAN

Information and communication technologies (ICTs) are emerging as powerful tools for development. They are vital for the long-term growth and efficiency of a nation. However, developing countries, which arguably need ICTs the most, face formidable obstacles and challenges in harnessing them widely for the common good. These barriers are in the form of poor infrastructure, inequitable access, absence of basic skills amongst large sections of the populace, lack of relevant content, scarcity of resources and so on. These can often seem insurmountable in the Indian context, as they are so deep-rooted and endemic. What is the relevance of ICTs for India's masses? What are the chances that people in remote and backward areas would benefit from ICTs? How can popular diffusion of these technologies be accelerated? What is the role of the State in their dissemination? What interventions are currently underway in this regard and have these been successful? What has been the impact of these technologies on rural productivity or on public service delivery? These are some of the crucial issues that must be addressed in utilizing the full potential of ICTs in India today.

ICTs are still a relatively elitist, urban phenomenon in India. Their impact on the masses is yet to be widely felt. However, a range of innovative experiments is currently underway as governments and service providers increasingly deploy ICT tools to reach the rural poor. Innovations range from community networks that deliver e-governance and e-commerce to public ICT kiosks offering low-cost Internet access and from popular communication and computing devices to ICT-enabled healthcare and education.

There is a range of noteworthy examples in rural ICT dissemination in different parts of the country. These include Gyandoot community network to deliver e-governance and other services to remote, tribal pockets in Dhar district of Madhya Pradesh; the M.S. Swaminathan Foundation's 'information villages' in Pondicherry, where villagers can access expert agricultural and other relevant information electronically; the Warana project in Maharashtra where ICT has transformed a rural cooperative; IIT Madras' Wireless in Local Loop (WLL) solution; and the SARI initiatives in Tamil Nadu for inexpensive communications.

Other ICT applications that are similarly transforming traditional interfaces between service providers and clients are mass e-governance projects such as *Bhoomi* in Karnataka, where land ownership information has been made freely available to citizens for the first time in independent India. The expansion of markets and portals to deliver rural products to a growing clientele, and the use of remote satellites to deliver health and education services to hitherto distant outlying areas, have brought in unprecedented changes in the rural settings. Not all of these initiatives are uniform in the intensity of technology deployed and the scale of their impact. Some are clearly local level projects, while others have the potential to be scaled up. Some are more sustainable than others. However, all these projects illustrate innovative applications of ICTs to improve the lives of the poor in unconventional ways, often in environments unfriendly to ICT induction.

A number of these initiatives highlight fundamental dilemmas of ICT induction. For instance, how does one go about estimating the technology needs of a populace? What level of technology induction would be deemed appropriate in a given context? What is the trade-off between ICT needs and other, more pressing, day-to-day concerns? Have ICT projects in fact delivered tangible, measurable gains to citizens, compared to costs? At the operational level, how does one tackle chronic issues such as access, content, language, etc.? What is the pan-Indian experience? What is a successful project, and what are its ingredients? What are the lessons to be learnt? These are some of the issues to be explored in the potentially high-impact ICT projects in the State of Kerala. Kerala offers an interesting framework to examine these issues, owing not merely to the State's advanced developmental attainments, in particular its high literacy levels, and consequently higher level of ICT preparedness, but also because of its pioneering and successful models of mass dissemination.[1]

The case studies cover FRIENDS (Fast, Reliable, Instant, Efficient, Network for Disbursement of Services), an example of a relatively simple ICT-enabled access point to government for improving service levels; the Information Kerala Mission (IKM), a pioneering e-governance project that promises to transform administration at the grassroots; and Akshaya, Kerala's pioneering ICT dissemination initiative. Each case study reflects a distinctive approach to ICT induction. Collectively, these cases highlight common dilemmas inherent in the endeavours and provide insights into how major challenges have been addressed.[2]

FRIENDS

FRIENDS and Information Kerala Mission represent two unorthodox approaches to ICT induction that promise radical change in the experience of governance for the citizen. Traditionally, back-end computerization has been the prime focus in government, rather than front end. Similarly, governments have typically followed a 'trickle down' approach in inducting ICT with higher levels being targeted first. Such trends have been reversed in these two projects. FRIENDS shows that the extent of success is not determined by the *intensity* of technology deployed but how *effectively* it is leveraged to improve the overall productivity in a system. Even fairly technological interventions, effectively leveraged, often yield dramatic results. This is important in the Indian context where resources, technology and management capabilities are usually scarce. FRIENDS is a rare instance of quick results in e-governance, which focused on citizen front-ends, rather than waiting for the completion of back-end re-engineering, which is often time consuming and problematic.

FRIENDS centres are single-window, IT-enabled, front-end facilities where citizens can remit payments to government and public utilities and access basic public services at one point. Besides the convenience of interacting with a single agency rather than multiple departments, these centres offer other advantages such as accessibility beyond office hours and on holidays with a modern ambience in which quick courteous service is offered by a trained staff to attend to customer needs. The centres also often function as help desks. Starting with one centre in 2000, 14 such centres are today operational – one in each district headquarters – catering to about five million citizens in the State. From modest beginnings these centres have quickly evolved into popular facilities and have serviced over half a million citizens so far. On an average, these centres service about a thousand customers a day. A vast majority visit the centres to remit taxes and pay utility bills and fees. Customers do not presently pay any addition fee for services at FRIENDS. However, the project plans to move to a self-sustaining revenue model in the next phase of its expansion.

The major success of these centres owes a great deal to consistently high levels of customer satisfaction. An independent World Bank study in 2002 rated the level of customer satisfaction at 99.7 per cent, as against less than 30 per cent with conventional government departments. The study estimated that the cost and time savings to citizens is significant; 80 per cent of customers have never had to wait to be serviced in FRIENDS, as against 61 per cent who invariably have to do so when served conventionally. In the remaining 20 per cent of cases, the average waiting time is only five minutes in FRIENDS, whereas it might be anywhere between 25 to 30 minutes at government departmental counters. Delays of over a day are not uncommon in government departments, but this does not occur in FRIENDS. The cost savings to the citizen are equally dramatic. On an average, citizens incur only one fourth of the cost to access services at FRIENDS as compared with other access points. Overall, the study found that 97 per cent of customers rated service levels at FRIENDS as superior.

Savings accruing to government departments was not similarly evaluated. It would not be insubstantial in terms of overall employee time saved, as well as cost per unit of revenue realised. However, while staff productivity and customer satisfaction levels indisputably scaled new heights, the project has definite shortcomings. Back-end integration with departmental processes continues to be poor. The project is seen as a laudable stand-alone initiative, but commands little ownership by mainstream departments. Centres have not evolved beyond the initial remittance-information offerings to a wider basket of services. Nevertheless, the project does offer an example of a low cost, high impact model, a robust 'hand pump' technology solution that successfully addresses the key challenges of e-governance in countries like India where compartmentalised governance, legal procedures, inadequate resources and time-consuming re- engineering can stall projects for years. The key lesson that FRIENDS ultimately imparts is that it is not the level of technology, but the ability to service customers in tangibly different ways that matters most, and that ICT can be a potent tool in this process.

Information Kerala Mission

Information Kerala Mission (IKM), the flagship e-governance project of the Government of Kerala, is an attempt to strengthen local self-governance through ICT applications. It envisages computerization and networking for the 1,215 local self-government institutions in Kerala. It is the largest and most comprehensive local computerisation project in India. It addresses the entire gamut of issues concerning local body governance, decentralized planning, and local economic development.

While FRIENDS illustrates how sharply focussed ICT intervention can perceptibly improve the interface between citizen and government, IKM demonstrates a strategy of radical change across institutions through extensive ICT induction. As observed earlier, e-governance efforts in India have so far mostly tended to focus on the higher levels of decision-making. Noteworthy efforts have indeed been made in innovative ICT induction in urban, local self-government institutions. However, a majority of citizens interact with grassroots governance institutions on a day-to-day basis. For a significant number of citizens government primarily means the local panchayat. Notwithstanding the increasing importance of local self-government institutions to a large number of institutions, inadequate attention has been paid to modernization and reforms at this level. IKM was conceived to address this important lacuna.

The Government of Kerala embarked on a radical decentralisation in the latter half of the 1990s, devolving sweeping powers and functions to local self-government institutions in the State, and entrusting these institutions with enormous responsibilities. The new role not only cast immediate and onerous financial and administrative responsibilities on these institutions, but also aroused higher public expectations for greater efficiency and transparency. The revised scenario also demanded a complete revamping of moribund systems and procedures to enable quicker responses for citizens.

The problems confronting local self-government institutions were manifold. Years of neglect had atrophied many legacy systems devised earlier. The staff was often corrupt and ill-equipped to deal with expanded responsibilities. Elected representatives were yet to come to terms with a vastly changed scenario. IKM was conceived as a comprehensive ICT intervention to address these challenges. The project involved a complete re-engineering of all key processes followed by comprehensive automation of finance, personnel, planning, welfare and developmental functions. The IKM project team worked closely for over three years with elected representatives, staff and community members to (1) overhaul and rationalise procedures for ICT induction, (2) devise appropriate decision support and management information systems for effective monitoring, (3) design and develop modern citizen-ICT interfaces, (4) automate workflows across departments, and (5) build ICT capacities to operate the new system.

IKM envisages a phased transformation of the existing systems to electronic systems. It has developed methodologies that suit this purpose. IKM has adopted a human-centred approach to e-governance. This approach is characterized by the holistic and proactive evaluation of existing systems and legacy systems, attempts to simplify and transform existing systems and effecting integration of systems. Systematic attempts at process reforms are also part of it. These would enable faster and objective decision-making, more citizen-friendly interfaces and better accountability. Special features that were incorporated included: (1) a comprehensive citizen portal where information relating to major developmental initiatives, entailments and grievance redressal could be available on a regular basis, (2) an information repository for elected representatives to track key parameters, (3) an integrated resource based information-system for local planning, (4) a comprehensive citizen's database, and (5) IT-enabled kiosk style disbursement of all major services and welfare entitlements.

IKM methodology places the employees and functionaries at centre stage in this transformation and focuses on their empowerment and capacity building as the mechanism for improving performance. The software applications are developed through active user participation. Emphasis is placed on demystification of technologies and establishing adequate technical support systems. Training and handholding are given high priority. IKM has taken out extensive pilot deployment of its application suites. With twelve software application suites ready for roll out, IKM is gearing up towards state level implementation in a Build-Transfer-Maintain (BTM) mode.

Owing to IKM's unprecedented scope and scale, the project had to surmount several technological and administrative challenges.[3] The re-engineering and software development efforts required for overhauling and automating the range of activities of 1,400 urban and rural local self government institutions, training of over 10,000 functionaries and canvassing of stakeholder support were huge challenges. Poor acceptability at political and bureaucratic levels in the initial stages, the unaccustomed rigour induced by ICT induction, the feeling of loss of control (predominantly in the bureaucracy), and apprehended loss of political and bureaucratic patronage were others.

The first pilot project for IKM was launched at Vellanad village in Thiruvananthapuram. Within one year the project reported all round improvement in revenues, service levels, employee productivity, transparency, participation and quality of decision making of the panchayat. The revenue to the panchayat in one single category (property taxes) increased three fold during this period, and is expected to increase further in the future, owing mainly to surveillance and monitoring.[4] The project has delivered the most significant results is in improved service levels. Vellanad has around 1,300 pensioners receiving welfare entitlements such as old age and disability, agricultural worker pensions, social assistance benefits, unemployment doles and so on. Earlier it would take a week to ten days for disbursing the pensions.[5] Pension disbursement is now done in a single day. While the quantum of the pension disbursed may not be huge in absolute terms, its value for those who constitute some of the poorest and most deprived sections of society cannot be overestimated.

Less dramatic, but equally important changes in other areas are also gradually visible. For example, the panchayat president and the standing committees now review all major activities on a weekly basis. Earlier, these reviews typically took place once a month. However, formidable challenges remain. IKM is a hugely ambitious venture, with all the issues inherent in change management projects. These include technical and logistical issues such as large-scale equipment procurement and deployment across 1,000 locations, extensive organizational restructuring, procedural reform across departments and institutions, management of transition problems and stakeholder expectations. By its very nature, IKM is a long-term undertaking. Its full impact will take years to unfold. One thing is certain, however: the project is set to mark a paradigm change in grassroots governance in India.

Akshaya

Akshaya is Kerala's pioneering ICT dissemination initiative launched in the northern district of Malappuram in November 2002. Conceived initially as a modest IT literacy drive covering one district, the programme subsequently evolved into a mass dissemination project aimed at universalising ICT access and skill development in the State. The project has three components that correspond to the central issues underlying the digital divide and constraining widespread ICT usage by the ordinary citizen – lack of access, skills and relevant content.

The project has proposed the establishment of an excessive network of public ICT kiosks across the State as self-employment ventures to address the issue of access, a mass e-literacy campaign to address the issue of skill development, and a major content creation initiative to provide relevant content. One Akshaya centre or kiosk would cover 1,500-2,000 families, that is, one centre every two to three kilometres in the State, on average. Funded by panchayats, the e-literacy programme would cover one person from each family. Typically, each Akshaya kiosk would have 5-10 computers with Internet access and at least one trained person to man the centre.

A key distinction between Akshaya and other similar ICT projects of its kind is its integrated or holistic strategy to address the three issues of access, skill and content simultaneously.[6] Most ICT projects hitherto implemented attempted to address one, or at the most two, of these issues at a time. For example, most kiosk models in the country have typically focused on the single issue of access, with little attention paid to content.[7] In Akshaya, on the other hand, creation of relevant popular content in the local language is the key to the whole endeavour. Effectively leveraged, the Akshaya network could eventually emerge as a large rural retail consortium.

It is therefore a unique endeavour to perceptibly enhance the overall level of ICT demand through a mass awareness and training campaign covering the entire State. The campaign, which is modelled along the lines of the earlier Kerala literacy programme, aims to mobilize demand not only by familiarizing the community with basic ICT tools, but, far more importantly, by educating the poor on its potential applications in their own lives. It is often assumed that lack of access constitutes the predominant constraint to widespread ICT diffusion, and that given easy access, people would automatically use ICTs. This view overlooks the important role that factors such as fear and inhibition of technology, perhaps the most challenging barriers, often play. This challenge is what Akshaya seeks to overcome, so that common people are empowered as comfortable users of technology.

The undertaking is huge in both ambition and scale. The first phase itself, covering the district of Malappuram, envisages over 500 Akshaya centres and basic ICT skill creation for 650,000 people. Based on the experience gained from Malappuram, the project is expected to create 5,000 retail ICT centres. These will be provided with skills to cater to approximately five to six million citizens across the State over the next three years. The challenges in a venture of this magnitude are enormous. Solutions must necessarily evolve as the project moves forward. An inevitable question that arises is: Can mass ICT induction on such a scale be sustained? Can ICT penetration, which is currently abysmally low in India at less than one percent of the population, be accelerated to over twenty percent within a brief span of two or three years? The substantive risk underlying Akshaya is the premise that a large network of ICT centres can succeed as financially viable entities when ICT demand can be sustained at adequately high levels once training is over.

Therefore, the future of Akshaya will be contingent upon the creation of demand for new ICT products, which in turn would depend upon service provider innovativeness in motivating the rural poor to use ICT tools to derive tangible benefits on day to day issues. Whatever the success rate amongst these centres, it is clear from the Malappuram experience that this experiment is triggering exciting and profound social changes. Malappuram is a conservative, backward and predominantly Muslim district. Yet, the project is drawing large numbers of women out of their houses and into ICT centres. Long queues of young and old women are a familiar sight at over 500 centres set up so far, and 50,000 people have been trained in the district. In remote villages, ICT centres are emerging as popular hubs of social life. The campaign is still in its early stages. It is far too soon to predict how information flowing into the community and the communication access available will impact life styles and relationships. However, it is clear that these will be transformed considerably by ICT induction.

Akshaya is one of the most ambitious ICT dissemination projects undertaken anywhere so far, because it seeks to empower an entire community across a State, including its most backward and remote segments. While FRIENDS represents the tried-and-tested approach towards using ICT for public welfare, where citizens are passive recipients of services, Akshaya represents a generational leap in its promise of providing powerful ICT tools to each household, for use in ways that would be relevant to their own requirements. Akshaya seeks to use ICTs as a powerful lever for enabling the poor and the disadvantaged to overcome traditional barriers that have stood between them and the unfettered exercise of their rights and choices for economic development, and for playing a more meaningful role in society. The manner in which the Akshaya Project unfolds, and its failures as well as successes, will provide potent lessons in understanding the efficacy of ICT in transforming ordinary lives, especially among the rural poor.

Notes

1. Kerala's campaign for universal literacy in the 1990s became the national model to improve literacy levels across the country. Akshaya, the ongoing mass ICT dissemination project, is the latest in this genre. A multidimensional project that aims to catalyse wide ranging social and economic change, Akshaya represents a unique experiment in mass ICT dissemination.
2. All three projects illustrate the potent role of ICT, as a catalyst and agent of change in modes of governance in projects such as FRIENDS and IKM, and in bringing about social and economic empowerment in initiatives such as Akshaya.

3. A uniquely innovative and participatory project development and management model, imaginative reengineering, intensive capacity building efforts at every stage of the project and enthusiastic championing by a section of young elected representatives are key factors that help surmount these challenges.
4. Earlier, over 50 percent of the properties were either unrecorded in the panchayat records or were in default. An extensive house-to-house survey conducted as part of the project resolved this problem. It has been conservatively estimated that the combined revenue of the local self-government institutions in Kerala would go up by over three times on this account alone, recouping the cost of the project in less than a year.
5. There have been instances when pensioners have had to visit the office two to three times, on average, just to check when their turn would be due.
6. Besides e-literacy training, Akshaya kiosks would offer a variety of ICT services including popular Internet browsing and email, besides value-added content such as web chat, VOIP, IT-enabled education and computer education, health, remote consultancy and e-commerce.
7. Consequently, a majority of these kiosks have been unable to convert initial enthusiasm for ICTs into sustainable, long-term demand. While this has been addressed to some extent in recent cyber café or web store models sponsored by Internet service and large content providers, the content offered is predominantly urban and its viability in rural areas remains suspect.

2

Akshaya

An Experiment in 'ICT For All'

T.K. MANZOOR

Like any other noble venture, it started as an inspiration. Then it got a character and a moral fibre. As it progressed, it took a definite shape. Finally, it became a reality. Akshaya has now been recognized worldwide as the largest community-based initiative in spreading e-literacy, empowering local community and employment generation. The lead came from the district panchayat (Village Council) of Malappuram, a 'backward district' that has huge workforce in the Gulf. Knowing that a basic knowledge of computers will enhance job prospects in the Gulf, they approached the government with a proposal for massive computer literacy campaign. The Kerala State Information Technology Mission thought it should be fine-tuned. They decided to study other ICT initiatives in the country and consult agencies such as the Science and Technology Entrepreneurship Development (STED) Project and C-DIT. That resulted in the birth of 'Akshaya', which would transform the lives of some 30 million people in India's most literate State.

'Akshaya' means perpetual prosperity. It aims at bridging the digital divide by disseminating e-literacy among citizens and providing sustainable ICT access points within the reach of the common person. This involves establishing around 3,000 broadband-enabled information hubs called Akshaya e-centres across the State. Each e-centre, run by private entrepreneurs, will cater to the requirements of around 2,000 to 2,500 families. They will provide locally relevant content and various services with the power of networking and connectivity to the common man. In the initial phase of the project, each Akshaya e-centre imparts basic skill sets to operate a computer to at least one person in all the 6.4 million families in the State, making it the largest rural e-literacy training project of the world.

Akshaya works on a Public-Private Participation (PPP) model. It also ensures significant involvement of local bodies at the grass-roots level. This project takes a multidisciplinary, developmental approach. It addresses issues in multiple disciplines of ICT for all, entrepreneurship development, building up infrastructure, the empowerment of women and so on. Akshaya was piloted in Malappuram district and its rollout in other districts is in progress.

Key Issues Addressed

There are three specific issues addressed by the project.

Access Points

To ensure ICT access points for all groups of people, this project aims at setting up at least 3000 Akshaya e-centres across the State. Each centre will be positioned such that it meets the requirement of 2,000 to 2,500 households in a particular area. Each Akshaya centre is to be equipped with 5-10 computers, a printer, scanner, web cam, other peripherals and broadband connectivity to carry out various ICT-based services.

Skill Sets

After the setting up of the Akshaya e-centre, the key focus is to impart e-literacy training to the people. This provides common people with basic skill sets to operate a computer, to use Internet facilities such as email, and also to create awareness about various applications of computer and Internet that can change daily life by saving time and reducing cost. The e-literacy training is a 15-hour structured programme developed in the local language. Colourful content is developed for this purpose with video clippings and digital images. Pedagogy includes traditional games for familiarizing participants with keyboards, mouse controls, etc. The training content is developed in the mother tongue, Malayalam. To cater to the needs of people living on the State's borders, e-literacy content is also developed in Tamil and Kannada as well as English. In this phase, e-literacy is being imparted to at least one person from each family, targeting 6.4 million e-literates in the State.

Content and Services

Are 'skill sets' and 'access points' sufficient to ensure ICT uses among the common people in Kerala? The question raised in the conceptualization of the project showed the way to address the issue of 'locally relevant content and services'. The content deals with locally relevant themes, structured in a manner that empowers users with information that will affect their lives in a positive way. It also removes the language barrier for getting information online, by developing the content in the local language. A wide range of topics include education, career development, agriculture, health, law and justice have been identified for developing content.

The Akshaya e-centres also intend to deliver value added Government to Citizen (G2C), Business to Citizen (B2C) etc. services to the people. The locations of the Akshaya e-centres are strategically planned and spatially distributed to cater to the people in the most remote areas of the State. They form a powerful network to bring the full benefits of e-governance initiatives of the government to the common man. E-payment, e-learning, e-business and e-governance are the priority sectors of services through Akshaya e-centres.

Distinct Features

Multidisciplinary and Integrated Approach

Instead of addressing single ICT development issues, care is being taken in Akshaya to ensure integration of key areas of issues for end-results. There are situations of ICT initiatives where access points may be created but people may not be imparted with the proper skill sets or given awareness in order to break psychological barriers and to make use of it. In other situations, where the skill set is imparted, the people may not have accessibility to relevant content that can be of some use in their daily life. In Akshaya, these issues are properly addressed by integrating the activities of creating access points, imparting skill sets and making available the locally relevant content and services to the people. Akshaya is basically an ICT project. However, it also touches issues in distinct disciplines like entrepreneurship development, expansion of infrastructure, capacity building, empowerment of women and so forth. It integrates developmental issues in distinct disciplines into a common platform and contributes in these areas to achieve specific goals.

Public-Private Partnership (PPP) Model

The project works on the Public-Private Partnership (PPP) model. Kerala has government-owned access points in the core sectors of development. Primary schools for education, primary health centres for medical care, Anganwadi for parental care, etc., are indicators of government-initiated development in the State. Deviating from this conventional approach, Akshaya has adopted the PPP Model, keeping in view better dynamism and effectiveness in the delivery of services to the people and also the long-term sustainability of the project.

Role of Local Governments

This initiative intends to make available the far-reaching benefits of the Information Technology (IT) for ordinary men and women in every nook and corner of society, by spreading its wings at the grassroots levels. To ensure that it reaches the grassroots level and, also, in the context of decentralized planning, the involvement of the local-self government institutions is most significant. In the 3-tier Panchayat Raj system, the State has 14 District Panchayats, 152 Block Panchayats and 991 Gram Panchayats. Kerala also has 54 Municipalities and five Corporations, which are local governments for urban areas. These three tiers of institutions function in tune with the concept of decentralized planning. Involvement of local bodies is found to be necessary to ensure grassroots level community involvement for the smooth and successful implementation of the project.

In Akshaya, the local bodies play a vital role from the beginning. The first step in this project is the finalization of the locations for the Akshaya e-centres. The locations are selected based on the state-wide spatial mapping done by the Town and Country Planning Department. Different locations are proposed by the Town Planning Department taking account parameters such as the number of households, accessibility, and proximity. The number of locations identified by the Town Planning Department was almost three times higher than the centres actually targeted. Out of these proposed locations, each local body selects the locations for Akshaya centres. The main criterion was the range of households between 2,000-2,500 within the catchments of each centre. This is an elimination process conducted by local bodies that finalized the outcome of exercises carried out by professionals.

After finalizing the locations, the next important step is the selection of the entrepreneurs to start the Akshaya centre in the proposed location. In the process of identification of entrepreneurs, promotional efforts include awareness workshops in each local body. Choices for submitting applications online or offline are given to potential entrepreneurs. On receiving applications, a primary screening process is carried out centrally at the State Project Office. Final selection is based on an interview in which the interview committee comprises mainly members of local bodies. Based on the preliminary screening, a short-list of potential candidates is prepared for interview and forwarded to the interview committees, which make the final selection of Akshaya entrepreneurs in their locality. After the selection of the entrepreneurs and the setting up of the Akshaya e-centre, there are continuous efforts to ensure involvement on the part of the local bodies during the e-literacy phase and afterwards.

Local Campaign

To ensure the reach of the project at the grassroots level, different types of campaign activities are also being organized. The campaign at the grassroots level starts with household visits. A group consists of the entrepreneur of the particular locality, elected representatives of the local bodies, and social workers who visit each and every household and convey the message of Akshaya to the people. Apart from the household campaign, local processions, meetings, and traditional art forms are being organized at locations where people participate actively.

For making the grassroots level activities more effective, a social animator is being deployed in each of the local bodies during the e-literacy phase. The social animators are nominated and selected by the local bodies and trained by the Akshaya project office. The social animator is the representative of the concerned local body where they are being positioned. A social animator works as a point of interaction between the public, entrepreneur, local body and the Akshaya Project office.

Entrepreneurship Development and Capacity Building

Being the critical component of the project to offer sustainability, due care is taken from the beginning in identification and selection of the right kind of person to start the Akshaya e-centre. Unlike conventional practices, the entrepreneurs are not offered any investment subsidies or grants. At the same time, support for revenue generation is extended based on an approach of rewarding the entrepreneur based on effort. Accordingly, in the primary phase, opportunities for baseline revenue are provided to the entrepreneurs by way of e-literacy training where the training fee is met by the local fund on per capita basis and government plays the role of a facilitator.

The group profile of 1,024 entrepreneurs selected to start Akshaya e-centres in the first phase is depicted below:

Table 1

Age-wise Distribution of the Entrepreneurs

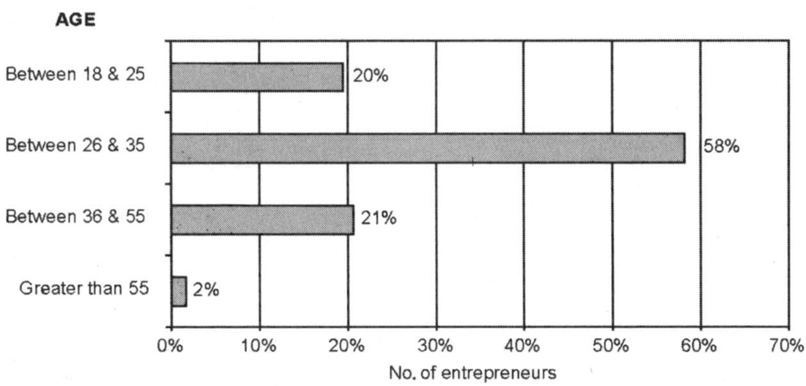

Seventy-eight per cent of entrepreneurs were below 35 years. The majority falls in the 'green' age bracket between 26 and 35. They are not too inexperienced or too old. As Table 2 reveals, 74 per cent of the entrepreneurs secured a degree or higher level of education.

It is heartening to note that the participation of women entrepreneurs is significant in Akshaya. However, whether local women

Table 2

Educational Profile of the Entrepreneurs

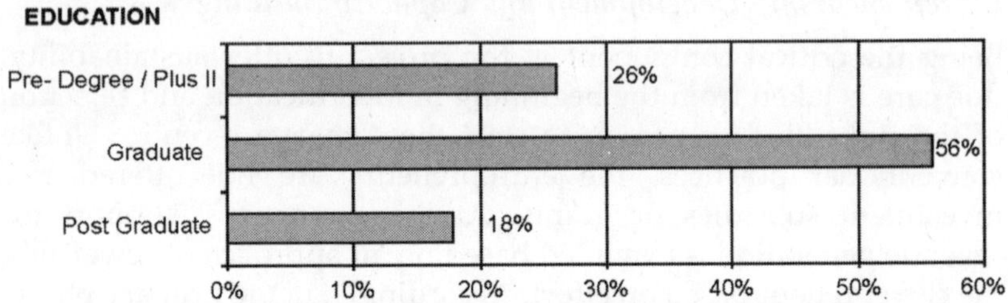

	Post Graduate	Graduate	Pre- Degree / Plus II
Education	18%	56%	26%

No. of entrepreneurs

Table 3

The Gender Distribution of Entrepreneurs

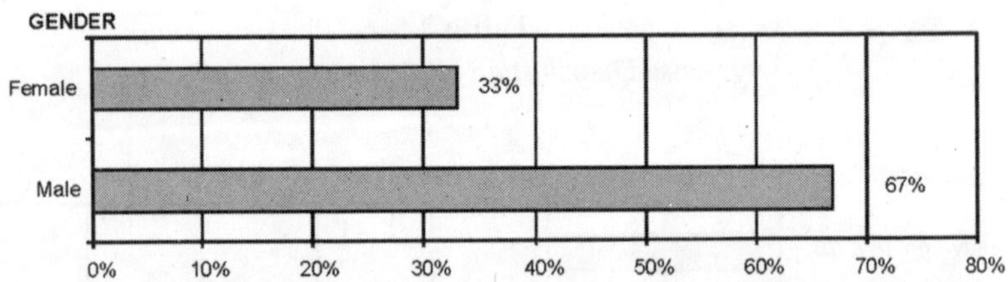

	Male	Female
Gender	67%	33%

No. of entrepreneurs

entrepreneurs will be more persistent and add sustainability to the project, is an important question that must be addressed.

Entrepreneurship development and capacity building are considered key areas of focus for this project. The entrepreneurs of Akshaya are provided with refresher and motivational packages from time to time to make them successful in the changing environment. The package is designed in modules of two to three days and imparted periodically once in two to three months, based on a brick-cement-brick approach. The training modules are designed with business/management games, role playing, and motivational laboratories. The group size is limited to 25 to 30 in a batch to ensure effectiveness of the interactive sessions. The brick-cement-brick approach provides more room for reinforcement of training inputs among entrepreneurs in the learning process. The entrepreneurs have enough opportunity to link the inputs to the practical entrepreneurial environment. Social responsibility of entrepreneurs is significant in the process of training and development. As in the case of entrepreneurs, the project makes constant efforts for continuous capacity building of stakeholders such as the representatives of the local bodies, master trainers, social animators and other stakeholders.

Promotion

Promotion is one of the main areas of importance in the implementation stage of the project. There are a number of promotional activities conducted at the state level and also at the grassroots level. Different forms of events and activities are accomplished locally to promote Akshaya. Seminars/workshops also are organized locally on topics like 'IT and Society', 'Akshaya in my Village' etc. School children take pledges in school assemblies. Street plays, nadan pattu (folk songs), monkey shows, and magic shows, among other activities, are used to mobilize people at the grassroots level. Traditional art forms also are performed that are relevant to the particular district/region. One of the major milestones in the mass campaign is the presence of Padmasree[1] Mammooty as the Goodwill Ambassador. Mammooty is a top rated cine artist who is techno-savvy. Such a promotional activity using a celebrity is used for effective branding of the project.

Women Empowerment

The project opens up immense opportunities to a large number of women in the State for self-employment or wage employment. Involvement of women is very high in this project at various levels. We can find their active presence in all the stages of this project. The greater response from women is due to the fact that the project creates opportunities in their own locality. There are a significant number of women who have come forward with a choice of an entrepreneurial career by starting an Akshaya e-centre. They are proud to be known as the women entrepreneurs of the Akshaya Project. The women entrepreneurs account for around 33 percent of the total entrepreneurs of the project selected in the first level.

Another area where women play a significant role in this project is as master trainer in the Akshaya e-centres. The master

Diagram 1

Percentage of Women Entrepreneurs

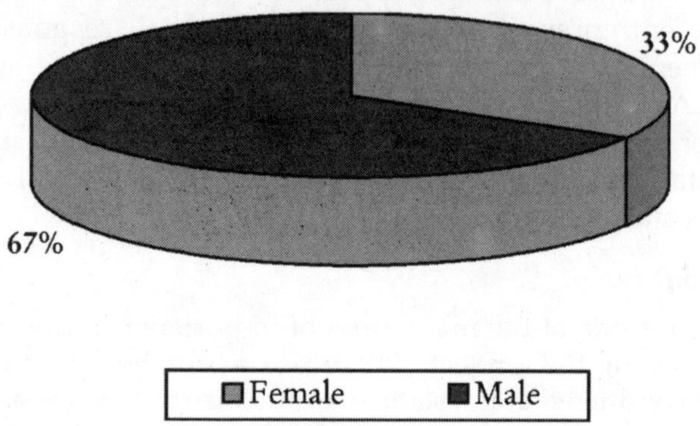

trainers are engaged by the Akshaya entrepreneur on salary basis. About 58 per cent of the master trainers engaged in Akshaya e-centres are women. The situation is similar for social animators. Women have come forward in large numbers to associate with this project as social animators who are co-coordinating panchayat level activities. In the State of Kerala, where a large majority of women are homemakers and the sector as a whole is considered to be an 'idle resource', the experience in Akshaya explicitly indicates the scope for empowering women by creating village level opportunities on a structured manner.

Training of Trainers

Another key aspect in the implementation of this project is to ensure the quality of the e-literacy training imparted to the common people. A master trainer is appointed in each Akshaya e-centre by the entrepreneur himself. Eligibility criteria are prescribed by the Akshaya project office. The entrepreneur remunerates the master trainer. The recognition of master trainer was based on a qualifying test. The qualified master trainers were imparted content specific training by the project office before starting e-literacy training. Around 1,000 trainers per district are deployed.

IT Expo

IT expo is one of the significant events that Akshaya organized to facilitate establishment of entrepreneurial ventures in a more effective way. This is a platform where the entrepreneurs and the vendors are able to interact with each other in order for equipment to be procured. The sole discretion for making the purchase decision remains with the entrepreneur. As part of IT expo, a pre-event technical session was organized for the entrepreneurs where they were informed about the specifications of the computers and the peripherals by the vendors. Eventually entrepreneurs bargain with vendors in groups and fix their prices. The IT expo facilitates procurement of equipment by entrepreneurs, but without loosing bulk purchase advantages. It was the first such event organized in the State of Kerala.

Recognition

As one of the biggest ICT projects of its kind in the world, Akshaya has received attention from people in different walks of life, inside as well as outside the country. Researchers, social activists, and students from different regions visit Akshaya and interact with the communities. The project has also received national and international awards.[2]

Akshaya Project has been selected for the Golden Nica award of Prix Ars Electronica in the 'Digital Community' category. Instituted by the Austria-based Ars Electronica, a platform for digital arts and media, the Ars Electronica Award consists of prize money and a 'Golden Nica' statuette, which are awarded annually in six categories: Computer Animation/Visual Effects, Interactive Art, Digital Music, Net Vision, Freestyle Computing, and Digital Community. The award is presented at a festival at Linz, Austria, in September 2005. The prize money was 10,000 Euros. There were 2,975 entries from 71 countries in the competition. Akshaya received the honour, as the project is conceptualized by integrating multiple objectives of mass e-literacy, entrepreneurship development, ICT interventions and community upliftment in the State. A seven-member jury selected the winners of six Golden Nicas, 12 Awards of Distinction and 73 Honorary Mentions. Ars Electronica said in a release, "Akshaya is one of the most ambitious development programmes ever launched to take advantage of information and communication technologies".

The Government of Kerala also bagged the India Tech Foundation Award – first runner up for the year 2005. This is awarded for becoming the most progressive state government in the application of IT in the State Administration (e-governance). Akshaya was projected as one of the e-governance initiatives during the contest. Akshaya also won Silver Icon Award in Innovative Operations and Best Practices – New Entrants for Exemplary Implementation of e-Governance Initiatives for the Year 2004. This is an award for implementing the project in Kerala to address the issue of the digital divide in the State in an integrated and holistic way. Providing ICT access to all sections of the society, the development of minimum skill sets to all the people through functional IT literacy training, and creation of local content to benefit all interest groups were the main components considered.

Critical Factors

The critical success factors of the Akshaya Project are economic sustainability of entrepreneurs and accessibility of the local community to ICT-enabled services. In the long run, the entrepreneurs are the key pillars of the project. E-centres run by entrepreneurs are to be effective delivery points for various services to the people. Support from other stakeholders is also needed for the long-term requirements of the project to ensure continued success and sustained developments.

Notwithstanding these factors, the success of the project in the pilot district of Malappuram has prompted the government to replicate the project in all other thirteen districts in the State. The rollout is currently in progress. The entire State is expected to be covered with e-centre access points even in the remotest rural part by the end of 2007. The project when completed will make Kerala a knowledge society and also a model of development for the entire world.

Notes

1. This is a special national honour conferred to those personalities who make significant contribution in their respective fields.
2. Only a few of those awards are mentioned here.

3

ICT-Enabled Information Systems in Developing Countries

A Case Study

K.P. SADASIVAN

Phase II of the World Summit on the Information Society, held in Tunis in November of 2005, was a landmark in ICT among the Third World countries. The convergence of experts on the assessment and evaluation of progress based on valid parameters was a sound exercise to understand the trends in this domain. The slogan 'ICT for All' carried the message to the developing and underdeveloped countries to review current trends and to formulate action plans for future.

The Indian scenario, during the last three decades, consequent to the unfolding of new technologies in the information transfer process, is perhaps a typical example of growth dynamics among the developing countries. It is well known that any major development contributing to the total progress of the society is always preceded by isolated attempts by individuals or individual institutions, to think, initiate and implement new ideas in the face of problems such as the scarcity of resources, conventional lethargy, rigidity, negativism or hostility that often prevail in the immediate environment. This is perhaps an inherent feature of the dynamics of society. The importance of this case study may be viewed in the backdrop of this context.

This paper is an attempt to focus on the developments taking place in the Library & Informatics Department of a major research and development (R&D) institute in India: the Regional Research Laboratory, Trivandrum (RRLT). RRLT is a national laboratory under the Council of Scientific and Industrial Research (CSIR) of India (URL: http: //w3rrlt.csir.res.in), located in Kerala, the southernmost state of the country. The thrust here is primarily on ICT-enabled systems and resource creation, with occasional mention of products and services.

The Library & Informatics Department of RRLT is a unique example of an 'early bird' in this region in the ICT-enabled information system development for three decades. The factors contributing to the development process include a visionary and supportive attitude of the top management, a reasonably fair receptiveness of research scientists to novel information and communication systems and services, the knowledge-base, initiatives and a proactive approach of the Library and Information System (LIS) personnel, exploratory ventures for financial resources, and supportive national policy decisions of CSIR to create infrastructural facilities for doing better R&D. However, LIS professionals play a pivotal role in optimizing these factors to create and deliver the best possible products and services to end-users. This is essentially the philosophy of success of LIS at RRLT.

The library activity was nucleated concurrently with the establishment of the Laboratory in 1976, with about 800 books, a few scores of periodicals, a modest annual budget allocation and a skeletal professional staff. A photocopying machine, an electronic typewriter, a microfilm reader, a slide projector, an overhead projector and an epidiascope were procured soon afterwards. The responsibility of organizing seminars and lectures – both internal and invited – was entrusted to the department as recognition of the concept that LIS shall be the centre of both scholarly information and communication within the Lab.

The major thrust during these formative stages was collection development covering books, periodicals, reprints, reports, standards, patents, databases and so forth. Gradually, a wide range of information services was introduced. This included traditional and state-of-the-art electronic information and literature access and acquisition facilities, information analysis and consolidation products for the research scientists, and an effective mechanism of resource sharing with local and national R&D and academic institutions.

CAN/SDI Service

During the late 1970s and early 1980s, the Laboratory was the first institution in Kerala to subscribe to bibliographic services under the CAN/SDI project from the erstwhile INSDOC, Madras. The goal was to assist various R&D groups through the provision of a Project-oriented Current Awareness Service. It envisaged regular supply of the latest literature from the computer tape version of Chemical Abstracts received by INSDOC under the National Experimental Project, funded by Canada. The regular receipt of the dot-matrix print-outs of SDI searches through the postal service by our scientists was really an exciting experience, as it was a unique opportunity to get the latest from Chemical Abstracts Service (CAS), the world's most comprehensive and most respected bibliographic database in chemistry. The service was a trendsetter for the scientific community throughout India. It must be understood that the current SciFinder or the multi-faceted web-enabled CAS services were all far beyond imaginary concepts at that point of time, when even a computer was not available for information processing!

A critical review of the activities during the mid-1980s prompted us to expand and diversify the programmes by adapting new systems and facilities that are emerging in the information – communication field across the world. I examine some of the significant facilities and services below.

Direct Online Search

Though the NASA-RECON Experimental Online Access Live Demo Programme from Frascati (Italy) to Bombay (India) was held in the 1970s, India remained without reliable commercial online access and search facilities from international databases. The major initiative was taken by Mr. N.V. Satyanarayana, the CEO of Messrs Informatics (India) Ltd., Bangalore. He played a unique and significant role in bringing the online and CD-ROM databases culture to India. This author had the opportunity to attend a DIALOG Online Search Awareness Programme held in Hyderabad in 1985. Inspired by the tremendous potential of the facility, a similar programme was held at RRLT in 1986 and was well attended by research scientists, academicians, students and LIS professionals from all over Kerala. Yet this was still the pre-computer stage. A telex terminal used for the search and connectivity was not satisfactory. Still, this was a major achievement for Kerala State and it opened up a new era of fast and up-to-date info/data accessibility from this remote corner of the world, then totally isolated from the available communication facilities in the country.

Online Search Facility of International Databases through DIALOG, California was established at RRLT on a regular basis in 1987. This was again the first of its kind in Kerala. In spite of the heavy cost of searches, it became a regular practice among the scientists to use the facility – searches were current or retrospective, bibliographic, patent, directory and market information, data for evaluation studies and so on. It became almost a culture of the Laboratory to depend on such database searches before initiating a project, for mid-course corrections for ongoing projects, for specific problem-solving issues, for disaster management decision-making processes, etc. In addition to indirect quality improvement of the R&D programmes, we had many success stories, including funding of new projects from major agencies. Awards for scientists and acclaim in international conferences, could also be achieved.

An important event occurred in 1993, when the Packet Assembler Dis-assembler (PAD) in Trivandrum in 1993 was commissioned by the Department of Telecommunications (DoT). Until then, the Gateway Packet Switching System (GPSS) was accessed through Bombay by STD line. The Trivandrum PAD substantialy lowered the telecom cost of online searches, and during the same period, the telex mode of access was shifted to computer mode. In 1994, the STN facility was added to the DIALOG service. Introduction of the Research and Education Network of the National Informatics Centre (RENNIC) in early 1995 added a new channel to access DIALOG/KR and STN through the Internet, supplementing the existing GPSS route via the PAD at Trivandrum.

The advent of CD-ROMs as an information exchange media in the 1990s directly contributed to the decline of online options. Currently, several publications and databases are available in the Laboratory in CD and floppy disk versions. CD-ROM databases available elsewhere in the country are also used occasionally for searches in resource-sharing mode by using the potential of e-mail channels.

Computer-Communication Network System

The first PC installed in our Library & Informatics Department in 1988-89 happened to be the second PC purchased for the Laboratory. Compared to the computers of today, the configuration was rather primitive. The RRLT library was the first among local S&T institutions to use a computer for creating in-house databases, using dBase II. Although we experienced occasional setbacks, growth was rapid thereafter. The use of computers in house-keeping operations, e-mail communication for purchase of documents such as a routine reminder service, and the regular analysis of productivity through the publications of RRLT scientists – all occurred earlier than other institutions in this region.

In August 1990, the installation of a User Node of the Scientific and Industrial Research Network (SIRNET) by INSDOC was a milestone in the development history of Computer-Communication-Network (CCN) system in the Laboratory. However, the Mail Node was in Madras and the connectivity was through dial-up mode using an STD line, which incurred a fairly high rate of tariff. Yet this was the first email facility commissioned in Kerala, bringing a new culture of communication through networks, to the Laboratory in particular, and to the region in general.

Attempts were then made to set up a Node of Education and Research Network (ERNET) in Trivandrum. RRLT showed great enthusiasm for creating the best possible system facility in data networking for the region by taking the lead role among the S&T institutions in Trivandrum. A high level meeting was convened in May 1993 (NETMEET-93) in the Laboratory. It was attended by Mr. Ramakrishnan, Director, ERNET project of the Department of Electronics (DOE), New Delhi, Dr. Remani, Director, National Centre for Software Technology (NCST), Bombay, Prof. T. Viswanathan, Director, INSDOC, New Delhi, Dr. A.D. Damodaran, Director, RRLT and senior representatives of the major S&T institutes in Trivandrum. Mr. T.N. Jayachandran, the then Secretary (Higher Education), Government of Kerala, informed RRLT that the Government of Kerala was also interested in joining the venture on condition that the universities be brought to the Net.

The primary objective of the project was to hook up Trivandrum with the ERNET hub at Bangalore or Madras. Mr. P.K.S. Menon, then Chief (Planning) of the Department of Telecommunications (DoT), was also associated and offered all possible technical help in establishing the communication link. But the project had to be shelved. DoT informed that the terrestrial-cum-microwave lines linking Trivandrum with both Bangalore and Madras would be completed only after another 12-18 months. Surprisingly, DOE had not proposed an alternative for the V-SAT link. Despite the failure of the project on purely technical reasons, the efforts stimulated the S&T institutes of the region to explore new communication network capabilities.

As part of the tie-up between NIC and CSIR at the national level, the Research and Education Network of National Informatics Centre (RENNIC) was commissioned at RRLT by hooking up through the NIC link at Trivandrum in dial-up mode. (The channel was subsequently made through the VSNL dial-up link through Madras). In January 1995, ERNET link was also created on a trial basis in dial-up mode with NCSI, Bombay. The official email facility was created in RENNIC domain. It also facilitated limited web, ftp & gopher search services using a Lynx browser. In 1997, a dial-up link with VSNL, Madras facilitated limited use of the Internet as an alternative. It is interesting to note that the bandwidth was so scarce that only 300 kbps was initially available, gradually progressing through the stages of 900, 1200, 2400 and 9600 kbps.

Another major development was the commissioning of CSIRNET, the VSAT-based network system with a bandwidth of 64 kbps. This was an exclusive network communication facility for CSIR labs with SCPC DAMA technology, with its hub in New Delhi, supported by exclusive transponder space residing initially in INSAT-2A and subsequently in INSAT-3C. This was operational in 1999/2000. Unlike similar institutions, email facilities were extensively available to users, including research students community. The goal was to widely distribute the benefit of the new mode of communication. At a time when the Internet had not reached the general public outside the Laboratory, this was a great boon to our users of all categories.

The Campus Wide Local Area Network (CWLAN) with 76 PC terminals interconnected, was commissioned shortly afterwards and hooked up to CSIRNET. With the usage traffic prevalent at the initial stages for email and Internet, about 30-40 users could concurrently access the Net at reasonable speed. As usage increased, there was a need for higher bandwidth. When BSNL started its network operations in Trivandrum with the introduction of the National Internet Backbone (NIB), RRLT was one of the early applicants for a leased line facility. Commencing with 64 kbps, the bandwidth was upgraded periodically, commensurate with our usage, to 256, then to 512 kbps and then to 1 mbps. Doubling of the bandwidth is now planned.

The LAN structure also saw major upgrades. Now the system consists of 260 computer terminals with 100 per cent coverage of the activity areas – both research and support wings – Internet access from the desktops of all scientists and staff, a central browsing facility and fibre optic links to external buildings. We have CAT5 (40%) and CAT6 (60%) UTP Cabling with Lucent and Panduit cables respectively as well as Cisco switches and routers, IBM Rack Servers and a host of other systems and devices. Our next upgrade aims at more connections, still higher bandwidth, new storage devices, better network security solutions and video conferencing to be implemented during 2006-07. Another major programme just initiated is the Management Information System (MIS), aimed at creating an effective Intranet facility. Internal Messaging System using Lotus Notes Domino is an added feature of Intranet service. The website of RRLT was launched in 2001 and is regularly updated. Exclusive websites for various international and national conferences and seminars organized by RRLT are also created and hosted.

CSIR Consortia

At present, the CSIR Consortium of e-Journals facilitates access to more than 3300 scholarly journals published by 11 leading publishers of the world. This Consortium, launched in 2002, was the first of its kind in a corporate sector in India. It facilitates unlimited access and download of full-text papers from the current volumes of these journals and back files of at least four years from the desktops of every scientist in IP-enabled mode. The CSIR Consortia initiative was formulated in the Fifth Meeting of the CSIR heads of Libraries & Information Centres held at RRLT in 2001. The author helped to initiate and implement the project by NISCAIR (CSIR), New Delhi. Even a year before the commencement of the CSIR Consortium operation in the 38 CSIR labs, four labs including RRLT had acquired the Science Direct access facility through a direct agreement with the Elsevier Publishers.

SciFinder, the powerful research tool provided by the Chemical Abstracts Service (CAS), Ohio, of the American Chemical Society primarily targets scientists in biochemical and chemical disciplines, is available in the laboratory. It facilitates access to current scientific information in journals and patent literature from around the world featuring options to search by chemical structure, biological sequence, substance name, reaction and research topic. RRLT played a lead role along with five other CSIR labs, viz., National Chemical Laboratory (Pune), National Institute of Oceanography (Goa), National Physical Laboratory (New Delhi), National Aerospace Laboratory (Bangalore) and Institute of Microbial Technology (Chandigarh) as the 'core group' in bringing changes in the LIS domain of all the CSIR laboratories.

Workshops and Conferences

A series of workshops and conferences on IT-enabled information systems and services were organized with the association of major national agencies with the objectives of promoting ICT-supported information transfer processes, educating end users on newly emerging trends in information products and services and developing human resources in information and knowledge management. The much-talked about Public-Private-Participation (PPP) of the present age in capacity building was actually practiced in our ventures even during yester years! Another remarkable event was the CSIR Golden Jubilee Invited Lecture (1991) delivered by Prof. F.W. Lancaster, University of Illinois (USA), the proponent of the 'Paperless Society'.

A project funded by the erstwhile National Information System for Science and Technology (NISSAT) of the Department of Scientific and Industrial Research (DSIR) India and successfully carried out during 2001-03 was another landmark. This involved a web-enabled information resource site on 'S&T News – India' as part of a national project entitled *Vigyan* (Knowledge). An offshoot was capacity building in undertaking more challenging projects of global nature.

The symbiosis of content and information technology has always been the hallmark of our LIS domain of the Laboratory. This was a consciously nurtured philosophy of information handling in the institution. This model has definite advantages over the alterative – a simple integration of the two concepts – in terms of value-added products and services, quality of deliverables, strength of information-user interfaces and transmission losses due to possible differences in perceptions. The fusion helps facile conversion of information to knowledge, by minimizing the skill component and maximizing intelligence.

The new breed of hybrid librarians, with expertise in both contents and IT infrastructure, is an explicit form of the above transition. Yet, the full spectrum of advantages will be visible only when the knowledge base of the subject-domain of LIS professionals improves. It is necessary to elicit the perceptions of users of their fast changing information needs and demands. It is necessary to evolve new models of information handling processes and data mining techniques, with a critical mass of refined and customized content embedded in IT as the end product. It is in the backdrop of the above scenario that the mandate of Library & Informatics Department of RRLT was aptly defined as: "The creation and management of globally competitive dynamic information–communication resource facilities by blending the normal and IT-oriented information systems with the objective of making information for action, ultimately enabling the scientists to do better and innovative R&D." We look forward to the co-operation and support of various national and international agencies and expert groups in creating and sustaining our dream project of a *Totally ICT-enabled One-Stop-Shop Infodrome* as a possible role model.

Barriers to Change

Resistance to Change

During the period when the laboratory was subscribing to three sections of the print versions of *Current Contents* a communication was received from ISI Philadelphia (now Thomson), the publisher. *Current Contents* was the unique advance contents pages information service covering the world's important journals in different disciplines. It was a highly popular weekly publication among the research scientists. The message was an invitation to switch over to the newly launched floppy version of *Current Contents*, viz., *Current Contents On Diskettes (CCOD).* This computer readable version had several new features when compared to its print version.

This author had the excitement of bringing yet another new tool for information search through CCOD. With several of its new features, one was sure that the user scientists would certainly opt for the same. I rushed to the then Director with the proposal. But he suggested me to seek the opinion of a senior Emeritus scientist working in the Laboratory, who was known to have considerable influence in various decision-making processes. His reply was: "We do not need any computer-based *Current Contents*; we are quite comfortable with its print version". I was terribly disappointed and had to shelve my ideas. It took another two more years for me to introduce CCOD – that too, as advised by the same Director, after having seen the features of CCOD operational in some other institution he had visited while on tour! I had mixed feelings. The system helped the users immensely for several years afterwards.

Attitudinal Aberrations

The same person who blocked our initial adoption of CCOD objected to the introduction of an email facility as well. This was at the moment that we were struggling to bring this novel communication system to our Laboratory and thereby to this region for the first time. His cryptic comment was: "We do not need these computer communication systems, which are sheer waste of money!" But this time I could win over our Director and the system was launched. The influence of Prof. T. Viswanathan, the then Director of INSDOC, New Delhi is thankfully remembered for his co-operation and support. It was the commissioning of the Scientific and Industrial Research Network (SIRNET) User Node. SIRNET was the subnet of ERNET. The Mail Node of SIRNET at INSDOC, Madras had to be accessed by STD call from Trivandrum. Despite the high cost of e-mail operations through SIRNET owing to the STD call charges, Dr. A.D. Damodaran, then Director of the Laboratory, encouraged the extensive use of SIRNET, saying that we are bringing state-of-the-art global communication systems to the scholarly community of the region. This is an example of how new ventures are discouraged by some and encouraged and promoted by some others in the same organization. Perceptions of individuals decide the destiny of institutions.

Positive and Visionary Thinking

The first DIALOG Online Search in the Laboratory was carried out in 1987 using Telex Terminal. The search was done on Derwent Patents Database and the topic of search was 'Rubberised Coir'. The search output showed that there was *only one patent* on the topic. It was a Japanese patent. The total search cost including the cost of the bibliographic information was around Rs. 3,500, which was considered as a prohibitively high cost! I feared that this would perhaps be the first and the last online information search, on account of the high cost involved for just a single reference (It was not even a full-text document) and that, too, with no comparison available. I decided to appraise of the situation with my Director. I was expecting the worst to happen. I was almost sure that I had to prematurely bury the great expectations of the new initiatives of online search of databases.

But, to my disbelief, my Director said not to bother about the cost. He felt we were bringing a new culture of information access to this region, despite its geographically isolated location from mainstream communication channels. The other inference from the search was that not much work has been carried out in the subject area and therefore our Laboratory could seriously initiate research in related topics. The episode revealed the positive and visionary approach of the CEO to facilitate changes happen.

Impact on Productivity and Collaboration

The quality and quantity of the productivity of publications of RRLT scientists has been steadily increasing year by year. The Library & Informatics Department has been regularly carrying out bibliometric/scientometric analysis of our publications by using tools like Web of Science, SCOPUS, and Journal Citation Report (JCR) (Science Edition). Our exercise has resulted in creating awareness among our scientists of the SCI Journals and Impact Factor Values. The message that we are spreading among our scientists community is not the old "Publish or Perish" slogan, but "Publish only in High Impact Journals, else in Non-SCI Journals". Our comparative studies with other CSIR Labs are based on SCI Journals.

With a total strength of 87 scientists, the Laboratory has been performing reasonably well. The productivity of publications during the last three years is indicative:

Year	*No. of SCI Papers*	*Average Impact Factor*
2003	107	1.979
2004	133	2.244
2005	146	2.382

The position of RRLT in 2005 among the 38 CSIR labs is:

Number of SCI Papers - Rank 4

Total Impact Factor - Rank 5

Average Impact Factor - Rank 6

RRLT has always been positioned above the CSIR average. A comparison with the other leading S&T institutes and higher learning centres for the same year, also gives interesting results:

Institution	*Average Impact Factor*
Tata Institute of Fundamental Research, Bombay	3.657
Indian Institute of Science, Bangalore	2.402
CSIR, India	1.967
RRL Trivandrum	2.382

It is significant to note that RRLT is fast achieving the status of Indian Institute of Science, Bangalore as per the above parametric scores. Although the above growth trends are the cumulative results of a series of factors such as the caliber of the scientists and the inspiring role of the top management, the direct and indirect persuasive role played by the Library & Informatics Department was quite significant. The activity group can legitimately claim to have played a participatory role through its ICT and Content Development programmes.

One of the events substantiating the role of Library & Informatics Department in initiating a collaborative programme was the relationship of the laboratory with Messrs Carborandum Universal in a project on aluminum abrasive materials. The preliminary discussions were amply supported by the relevant information inputs generated through extensive online search operations. Capacity building in this area could be possible primarily through the customized information gathered from the global resources through the then prevailing state-of-the-art tools and techniques.

Success Stories

Scores of stories on the success of ICT and the contents can be narrated. I will narrate one episode here. A central government team visited Kerala to discuss matters related to landslides occurring in the high ranges of Kerala State. The Government of Kerala had constituted a committee in which Dr. A.D. Damodaran played a major role. As usual, the demand for relevant and accurate information on similar occurrences came to us. The time available was just a few hours. The best possible and economic option then was to search a bibliographic database like Georef. The CD version of Georef was available at National Centre for Science Information (NCSI), Bangalore. Since Dr. T.B. Rajasekhar, Head of NCSI was a personal friend of mine, I sent a search query by e-mail and requested him to do the search on the CD-ROM database and send me the output back by e-mail. With current communication technology, these events may not seem novel, but then it was unique and exciting.

Regular interactions with the scientists on the usage and the needs of specialized information products and services and occasional feedbacks have always been part of our strategy. Manfred Kochen's concept of 'Information for Action' had a strong influence in moulding my professional philosophy. Productive scientists compliment the efforts made by Library & Informatics Department from time to time. The e-Journals access facility consisting of full text access to about 4,000 journals from the world's 13 leading publishers has made their research life very comfortable. Desktops of every scientist have Internet and e-journals access facility. Formerly, we used to arrange regular trips to Vikram Sarabhai Space Centre (VSSC), Trivandrum to use their technical library facilities. The service was operational for more than ten years. But ever since the e-journals access facility was introduced at RRLT, there is no demand for such trips and the service facility had a 'natural death'. Similarly, submission of papers for publication, pre-publication access to papers after refereeing and discussion forums on the Net are just a few examples of the privileges enjoyed by our scientists.

Yet another major advantage harvested by our scientists was through the extensive use of e-mail facility. In addition to the routine communications, it has helped enhance confidence and visibility among their peers in the respective fields. To quote an example, when I asked one of the productive scientists about the benefits of e-mail usage over a period of time, he replied:

> In the early days, when e-mail was not available, our communication on scientific matters was confined to our fellow professionals working elsewhere, whom we knew each other. But the e-mail opened up unlimited scope to discuss with peers the world over, whether you personally know them or not.... It really helped to enhance your knowledge base and better acceptance of your expertise; it broke the barriers of level difference between the peers and the working scientists...

This demystification of the communication barriers is a major development in all fields of science and technology. I am sure this paradigmatic change has helped to improve performance efficiency and excellence of all specialists who productively used such facilities. The above observation of one of our scientists is evidence of the impact of ICT-enabled global change on the Laboratory.

Nevertheless, technical difficulties and bureaucratic hurdles figured almost always in one form or the other. Manpower problems were so acute that one had to venture to do major works with the help of ad hoc recruits at fairly junior levels. Their commitments and contributions had greatly helped in building up major facilities. Our team was always an exception in the so called 'normal office time' work culture. Often I was feeling guilty of overloading my young colleagues, mercilessly snatching away even their precious holidays, and, of course, mine as well. My similar partnership was extended to a few ICT experts in other selected institutions and some vendors.

Bureaucracy caused delay in implementing certain programmes. Equally or more damaging were attitudinal problems from some key functionaries, consciously and/or out of ignorance. There were instances of even resigning the job and running away. But somehow, the strength derived from some kind of a determination to achieve certain goals enabled us to achieve what the Laboratory has now. What has been achieved may be great, but what remains to be achieved is endless. No doubt, development is a dynamic continuum.

4

Free Software in Kerala

V. SASI KUMAR

A friend who had worked in Brazil for a couple of years once told me, "Kerala is known in Latin America for Free Software." This indicates the extent to which Kerala has dominated the Free Software scenario in India. It is not by chance that the headquarters of the Free Software Foundation of India happened to be situated at Thiruvananthapuram in Kerala. The State is now poised to become the first in the country to introduce exclusively Free Software for IT education in high schools. We shall examine here how all this came about. But before that, we shall look at what the term Free Software means.[1]

Free Software is software that gives users freedom – four freedoms, to be precise. As the website of the Free Software Foundation (http: //www.gnu.org) says: 'Free Software' is a matter of liberty, not price. To understand the concept, you should think of 'free' as in 'free speech', not as in 'free ice cream'. Free Software is a matter of the users' freedom to run, copy, distribute, study, change and improve the software. More precisely, it refers to four kinds of freedom, for the users of the software:

1. The freedom to run the programme, for any purpose.
2. The freedom to study how the programme works, and adapt it to your needs. Access to the source code is a precondition for this.

3. The freedom to redistribute copies so you can help your neighbour.
4. The freedom to improve the programme, and release your improvements to the public, so that the whole community benefits. Access to the source code is a precondition for this.

Free Software began to be developed when Richard M. Stallman, a programmer with the Artificial Intelligence Laboratory of the Massachusetts Institute of Technology, resigned and started the GNU project in 1984. Later, he also started the Free Software Foundation (FSF). Today, there is a large volume of Free Software, and the most popular Free operating system is GNU/Linux (sometimes called simply Linux).[2] We shall now take a look at the story of Free Software in Kerala.[3]

The Beginning

The story began with the introduction of TeX, the typesetting programme that was designed in the 1970s by Donald Knuth, the author of *The Art of Computer Programming,* a four-volume classic. Prof. K.S.S. Namburipad of the Department of Mathematics in the University of Kerala introduced TeX into Kerala. TeX could typeset mathematical equations very neatly, which no other software could do, especially in the 1980s when Prof. Namburipad brought TeX in fourteen floppy disks from the United States. He could bring the programme and use it on a number of computers without any legal problem because it had no licences – it was in the public domain. TeX was, in a sense, the grandmother of Free Software, as some people call it.

Prof. Namburipad encouraged his students to learn and use TeX, especially for preparing their theses. One of his students was E. Krishnan, now with the Mathematics Department of the University College, Thiruvananthapuram, a leading exponent of TeX and one of the authors of the very popular *LaTeX Primer*[4] published as an electronic book by the Indian TeX User Group. Dr. Krishnan also played an important role in establishing the Free Software Foundation of India. Another person inspired by Prof. Namburipad was C.V. Radhakrishnan who once ran a small centre that prepared theses for the research students of the Kariavattom campus of the University of Kerala.

C.V. Radhakrishnan took a serious interest in TeX. He found that there was a business opportunity here with virtually no competition. Eventually, he established a company in 1995, called River Valley Technologies, for typesetting of scientific papers and theses. He had his two brothers with him when he started the company. They used the DOS operating system running on Intel AT machines, along with Novell Netware for networking. Since Unix was rather expensive those days, they did not attempt to use it, though they were familiar with it. Around 1996, a computer vendor who supplied part of their systems suggested that they use Linux, which was very similar to Unix and was free. He also gave them a CD containing the Slackware distribution. Around March, 1996, the magazine *PC Quest* brought out CDs containing the Slackware distribution of GNU/Linux, the first commercial distribution of the free operating system.

It was in all likelihood a copy of this CD that Radhakrishnan got from the vendor. Though the operating system was primitive in some ways, and installing it on a computer was a tough job, it came in handy for the new company. As Radhakrishnan says, "It was very difficult to instal Linux those days. It took us about one week to install it on one system. We could link it to Novell Netware since there was a tool for that. Later, we installed it on all machines and discarded Novell Netware." River Valley Technologies thus might have been the first Free Software based company in the State and almost certainly the first TeX based company in the country. Since then, the company has been using GNU/Linux almost exclusively, except for one computer that still runs MS Windows. This machine is kept mainly for opening MS Word files and for editing of vector graphics for which free alternatives are not yet sufficiently powerful. Much of their work is automated so that human intervention is required only minimally. Today, Radhakrishnan is one of the top TeX programmers in the world. His company has seventy employees with clients such as the Institute of Physics, UK, Macmillan (*Nature*) and Elsevier. The company uses Free Software and also sponsors India's first portal to host Free Software projects.[5]

Meanwhile, Satish Babu, CEO of the South Indian Federation of Fishermen's Societies (SIFFS), was using the new technology of the Internet to enhance the efficiency of his organization. He learned of the Internet when he went to the Hull International Fisheries Institute (UK) in 1993 for training in fisheries management, and became interested in the technology. Though basically a management expert, trained at the prestigious Institute of Rural

Management, Anand, his interest in the technology prompted him to study computers and programming in depth and virtually made him a programmer. Today, he is involved in running a software company[6] in the Technopark at Thiruvananthapuram and he is actively involved with bodies like the Computer Society of India and the Indian Chapter of the Institution of Electrical and Electronics Engineers. He is the Executive Secretary of the Society for Promotion of Alternative Computing and Employment (SPACE), NGO promoting Free Software, and an active member of FSF India.

After his return from the UK, Satish Babu began using computers to improve the efficiency of the organisation he was working for. He learned that email was being implemented through the Ernet network, which used VSAT technology at that time. Leo Fernandez of the Indian Social Institute, Bangalore, with which Satish already had links, helped him to link to the Ernet node in Bangalore in 1994 through a telephone dial-up connection. Satish configured his computer to dial up in the early morning hours when telephone rates were lowest, to send and receive mails. Though this was slow by today's standards, since an exchange of mails would take at least two days, it was much faster than using the postal service. Part of the software he used for this was actually Free Software, though he was not aware of it at that time. It was again Leo Fernandez who introduced him to Linux in 1995. Though not very confident about the new system, Satish and his friends soon grew to like it, especially since there were a lot of options for control at the system level. As Satish says, "once you learn to tinker around with the system, you really start enjoying it and it becomes a habit that is difficult to get over", GNU/Linux offered plenty of opportunities for such people. But Satish and friends were still largely unaware of the ideology of Free Software and its implications. Leo Fernandez gave his first distribution of GNU/Linux, known as Slackware, to Satish. Soon, SIFFS organised a training programme in GNU/Linux by Leo. In 1998, Radhakrishnan, Krishnan, Namburipad, Satish and others decided to set up a Linux User Group (LUG) in Thiruvananthapuram. Others associated with this included journalist K.G. Kumar, computer science student M. Arun, and P.M. Sasi, of the Centre for Development of Advanced Computing (CDAC), Thiruvananthapuram.

M. Arun played a special role in these developments. He was a student of Computer Science at the College of Engineering,

Thiruvananthapuram, when his father bought a computer from Keltron (in fact, the first computer Keltron sold) with an operating system developed by IBM known as OS2. As Arun says, "This was much better than Windows 95, which was popular at that time." One of his friends had a Packard-Bell computer with MS Windows from the US and wanted to instal Linux in it. Arun had a copy of the *PC Quest* Slackware CD and undertook the job of installing it in the machine. It installed neatly without any problem. So he decided to try it on his computer. But, the result was disappointing. He could not get the graphical interface running. One of his friends suggested that he try the Red Hat distribution, and Arun wrote to a company, GTL Enterprises in Bangalore, for a copy of the Red Hat CD. They replied asking him to contact the local Linux User Group that C.V. Radhakrishnan and others had just started. Arun went to Radhakrishnan's office with his friend Amit and not only obtained the CD but also joined the LUG for their weekly meetings.

Arun found that a few teachers in their college were interested in GNU/Linux. They asked the college to purchase some manuals from FSF, Boston. Arun and a few of his friends had read about the ideology of Free Software and were attracted by it. In 1999, Wros Publishers organized a conference called *Bang Linux* in Bangalore that Arun and other students attended. There they heard Richard M. Stallman – usually known by his initials RMS – the founder of the GNU project and the Free Software Foundation. This was his first meeting in India, but his lecture impressed the students from Kerala. They came back thoroughly convinced about the ideology of Free Software. Today, Arun is the secretary of the Free Software Foundation of India and is also the co-coordinator of SPACE.

With the new millennium came the group known as *Free Developers*. This was started by one Tony Stanco, an advocate from the US, who proposed that they start a company that would do business using Free Software and eventually make it the leading software globally. He had corresponded with RMS about his ideas. Though RMS was skeptical about the feasibility of the project, Tony went ahead with it and managed to obtain support from a number of people, including C.V. Radhakrishnan, Arun and others in Kerala. This initiative helped in developing the dotGNU project, which was a free substitute for Microsoft's. NET, since several people from India joined the project.

The discussions in the Thiruvananthapuram LUG soon led to the adoption of the ideology of Free Software, which the members found attractive. They discussed the idea of establishing a Free Software Foundation of India, and a unit of Free Developers. FSF India, they hoped, would act to supervise the ethical aspects of Free Developers. Radhakrishnan got in touch with Richard Stallman and got his approval for starting FSF India, and got him to agree to inaugurate FSF India. Satish Babu, then Regional Vice-President of the Computer Society of India, organised the inaugural function, which also served to inaugurate the Indian branch of Free Developers.

Freedom First

The conference organized at Thiruvananthapuram in connection with the inauguration of Free Software Foundation of India was aptly called Freedom First. The journalist, K.G. Kumar, suggested the name and it immediately struck a chord with the others. Richard Stallman was invited for the conference. He was received with honour as a state guest by government officials to discuss the philosophy behind the movement.

The organizing committee formed to conduct the conference included Radhakrishnan, Satish Babu, Arun, Krishnan, Rajkumar (who runs a Free Software business), P.M. Sasi, K.G. Kumar, Dr. K.R. Srivathsan (Director, Indian Institute of Information Technology and Management, Kerala – IIITMK) and others. During the public function, Stallman inaugurated FSF India, and the Secretary to the Government, Information Technology Department, chaired it. About 300 people were present, filling the auditorium beyond its capacity. "Computer users in India, as else where, deserve the freedom to share and change software, the way cooks share and change recipes. So I am pleased to inaugurate the Free Software Foundation of India, which will promote the use and the development of Free Software in this country", Stallman told the gathering. Later, he met the Minister for Information Technology and held discussions on promoting Free Software in the State. The afternoon session was devoted to the inauguration of Free Developers India and some technical presentations.

There were a couple of interesting moments during Stallman's visit. At the airport, a number of people were curious to see a figure with long hair and long beard and wanted to know who he was.

When he understood what they were asking, RMS introduced himself, "I am St. Ignatius of the church of *Emacs*.[7] Possibly, some people took that seriously! There was a poignant moment when Stallman was going to a hotel for lunch along with a few other people. One of them told Stallman that Nelson Mandela had signed a Freedom Declaration that had been put up at the Free Developers website. "RMS just couldn't believe that and he almost cried. He said Mandela had always been his hero," wrote Ramakrishnan. When someone tried to compare Stallman with Mandela, RMS retorted that whatever he has done could never be compared with the 25 years in prison that Mandela had suffered.

It was an achievement of FS enthusiasts in the State that the government agreed to support the event and treat RMS as a state guest. As a report in *Linux Today* said:[8]

> Government officials and other Free Software supporters in the State of Kerala believe that Free Software meshes particularly well with Kerala's long tradition of democracy, equity and public action. Just as Kerala is often held up as a model of equitable social and human development in the region, Free Software supporters there believe they can leverage the inherent freedoms of Free Software to evolve an equitable Knowledge Society based on software independence and self-reliance.

Advocacy

The conference was a great success in many ways. It attracted a great deal of media attention and popularized 'Free Software' and Richard Stallman among the public. There have been a few instances where the Free Software community was able to influence decision makers to choose Free Software over proprietary. In some cases the decisive breakthrough was achieved by individual effort, while in some cases, it was a community effort. We shall look into two cases here, that of the implementation of a network by the Public Works Department (PWD) and that of the introduction of IT education in schools.

One of the first successful campaigns for Free Software was in the Public Works Department of the State. InApp Technologies, the company started by Satish Babu, Amarnath Raja and others, was asked to make a proposal for a PWD project by one of the secretaries of the department. InApp made it clear that while they do work with all technologies, they would quote only for a free platform, as they considered it as most appropriate for any e-governance project.

The consultant to the Kerala Transportation Project, under which the application was being planned, felt that he did not know sufficiently about Free Software, and obtained quotes from Microsoft and Oracle. However, the Secretary concerned knew about InApp and suggested to the Principal Secretary that InApp's proposal should be considered seriously. A debate was therefore organized to (a) explain what Free Software was, and (b) what its advantages were over proprietary platforms. This debate was conducted by the then PWD Principal Secretary and was attended by two other secretaries, the consultant to the Transportation Project, several Chief Engineers and senior people from the PWD.

Two people from Microsoft, one from Oracle, and Amarnath Raja and Satish gave presentations. Satish was armed with a survey conducted among Technopark companies about their perceptions on Free Software in December 2002 (http://gtechindia.org/jsp/survey.pdf), which was conducted by Satish, the GTech Treasurer. The study had clearly shown the features of Free Software that made it attractive especially for e-governance projects. Microsoft came prepared with their 'sponsored research' findings. Since Satish and Amarnath were prepared for it, they were easily able to refute their findings. Microsoft may not have expected this, and perhaps had thought that this would be a walkover. The Oracle representative was unaware of the local politics and looked surprised by the ferocity of the debate. He started off in favour of Microsoft, but shifted to the Free Software camp halfway through. The Secretaries were convinced about the need to go Free, and InApp got the order. The application was delivered and is running well.

Another successful campaign, driven by a large number of Free Software enthusiasts with much more publicity was the drive for the inclusion of Free Software in IT education in schools. The Department of Education, Government of Kerala, started a project called IT@School[9] for bringing IT-enabled education to the high schools in the State. The project constituted a committee headed by Prof. U.R. Rao, former Chairman of the Indian Space Research Organization, to make recommendations on the direction the project should take. After studying the status of education in the State, the committee recommended that the project should aim at bringing IT to high schools in the State to empower teachers and to use the technology for improving curriculum transaction in the classes. However, the project found that most of the teachers and the students possessed few IT skills. They, therefore, decided to start IT education at the high school level initially. IT was thus introduced in the eighth standard in the year 2002 after conducting training in IT for a large number of teachers. The teacher training was organised with assistance from the Intel *Teach to the Future* programme. Their course material, wholly based on Microsoft software, was used for the training.

The textbook for IT prepared by the State Council for Educational Research and Training (SCERT) was based exclusively on Microsoft Windows and other Microsoft applications such as MS Office. The Free Software community in the State found this very offensive, since it ignored the existence of Free Software and promoted the products of one company, ignoring even other proprietary software. The community responded by talking to people, sending letters, writing in the media and so on. The Free Software User Group in Kochi prepared a memorandum and sent it to several people involved in the matter, including the Directors of the IT@School Project and SCERT, the Director of Public Instruction, the Principal Secretary, Education Department, and the Secretary, IT Department. They pointed out that:

- IT@School was promoting the software of one company at the expense of software produced by everyone else.
- The government would have to pay an enormous amount for licensing the software for the schools.
- Even if the company gives the software for the schools free of cost, it is only a marketing ploy in order to reap benefits of having a pool of people who are familiar with their software packages and thus form an assured customer base, either as users themselves or as potential skilled employees.
- The government's approach would result in compelling not only schools, but also the general public to purchase software from this particular vendor in the future. This would create a monopoly in favour of that corporation and expose the public, the State and the nation to the mercy of a single company.
- The Corporation, whose brands and products are prescribed in the syllabus, does not publish the standards used in their software. This practice compels others who interact with users

of the products of this corporation (such as the government and schools) to purchase software from this particular vendor only – a situation known as 'vendor lock-in'.

- The government is promoting illegal copying and installation of software in the computers in the schools by not providing for software costs.
- Handling-licensing issues is not simple and there has been at least one instance in which a school in the US had to pay $300,000 as a fine – even screenshots used in textbooks may have to be licensed.
- Several software packages, both applications as well as operating systems, which conform to industry-wide standards, adopted and maintained by independent vendors, and with less restrictive licenses, are available.

The Kerala School Teachers Association decided to throw its weight behind the demand from Free Software enthusiasts. The government and the IT@School Project were still not willing to change. However, due to pressure from several directions, SCERT decided to incorporate Free Software also in the textbook and rewrote the textbook for the eighth standard for the academic year 2003-04. Some schools, especially in northern Kerala, where there were teachers who knew how to install and use GNU/Linux, installed it and started IT classes using it. In order to ensure that all schools did buy sufficient computers and taught IT, it was also decided to include IT as an additional subject and conduct examinations, though with less marks than other subjects.

Computerized Examination

An important phase of Free Software started when the IT@School Project decided to conduct part of the IT examination using software. They developed software called *Softexam* for conducting the examination. This was designed primarily for the MS Windows platform and some of the schools using GNU/Linux had to install MS Windows to enable the software. There was immediate protest from the Teachers Association and the project was forced to develop *Softexam* for GNU/Linux also. However, this is now being virtually discontinued, with the software being confined to presenting previously prepared questions randomly and saving the responses for later evaluation by teachers.

One problem with using GNU/Linux was that there were several distributions of the OS, each slightly different from the others, and schools had installed different distributions. Even preparing the textbook became difficult, since the screenshots, and sometimes even the procedures for using the software, could be different for different distributions. To solve this problem, the Free Software Foundation of India suggested developing a custom distribution for IT@School, and eventually created the distribution with funding from the Kerala State IT.

Another problem faced by IT@School was that of providing support to the schools where GNU/Linux was being used. They called for private agencies who were willing to provide support to register with them. A number of agencies, including Free Software User Groups, responded and about twenty of them were short-listed. A final solution to the problem came when SPACE decided to offer support to IT@School, both in terms of updating the distribution used in schools and in providing support to the teacher community. The website of SPACE now has provided for teachers to post questions there, to which experts will respond, and also a page listing the Frequently Asked Questions and the answers to them. The IT@School Project arranged for teachers to be trained in GNU/Linux and a majority of teachers have already been trained. A Resource Centre has been established in Kochi for conducting teacher training with technical assistance from SPACE.

The Society for Promotion of Alternative Computing and Employment (SPACE) is a society promoted by Kerala State IT Mission with the objective of promoting alternative computing, that is, Free, Libre and Open Source Software (FLOSS). It has a government nominee in its Board of Directors and has support from such groups as IEEE, the Computer Society of India and academia. It had as its first Chairman the then Vice Chancellor of the University of Kerala, Dr. B. Ekbal, with members such as Satish Babu, Amarnath Raja, P.M. Sasi and C.V. Radhakrishnan. Though it came into existence in 2003, it initially had few resources and activities. This problem was solved when SPACE entered a relationship with SOMA for working on joint projects. SPACE obtained an office and recruited staff. Activities started in areas like promoting Virtual Micro Enterprises (VMEs) based on Free Software, advocating Free Software in colleges, setting up Free Software Cells, and helping to train school teachers in FS.

Some of the achievements of SPACE did attract considerable media attention. One example is the development of a distribution of GNU/Linux specifically for electronics laboratories in engineering colleges. This was made available on one CD, and was named *Free Electron*. The FS Cell in an engineering college in Thiruvananthapuram created the distribution with help from SPACE. There were a number of requests for the CD from colleges inside and outside the State, and even one from abroad. Another distribution that SPACE created for system recovery purposes was distributed by a local IT magazine. SPACE conducted a workshop in collaboration with Mediact, an NGO involved in media education. Held at a village library, this activity sought to create and publish a village newsletter in Malayalam, attracting much interest. Another programme that became very popular was the initiative for setting up a radio station for fisher folk. Initiated by a few young people from the fishing community in Thiruvananthapuram, Radio *Alakal* ('waves') could not begin regular broadcasting due to licensing issues, but started narrowcasting using loud speakers at specific locations. All the work for Radio *Alakal* was done using Free Software. SPACE also helped the IT@School Project to set up a teacher-training centre in Kochi.

In 2005, the government announced that schools in Kerala would make the transition to Free Software in stages. Supplements to textbooks were created to enable students to study using GNU/Linux and introducing software to children. Tuxpaint, simple painting software, which a child could use even without familiarity with the intricacies of saving or retrieving a file, and Gcompris, a set of games for learning how to use the mouse and keyboard, became very popular with children. The textbooks for all three classes in high school are now being revised to contain Free Software exclusively. Kerala is poised to become the first State in the country to use exclusively Free Software in its schools. It is also poised to become the first State to introduce and promote IT-enabled education in high schools.

Developmental Efforts

A certain amount of Free Software development was done in the State even in the initial days. This includes localization of the GNOME[10] Desktop, thus making the desktop available in the local language, a project monitoring application for the government and a portal for enhancing transparency in everyday activities of the government. Arun and his friends Gopal, Sreekrishna and others soon after the establishment of FSF India, started the localization work. The Managing Director of Keltron, an undertaking of the Government of Kerala, offered to help them in their effort. The idea was to provide the government and other users a platform in Malayalam for various uses. They received support from Asia-Pacific Development Information Programme (APDIP), an initiative developed and funded by the United Nations Development Programme (UNDP). The project sought to create Free Font, a toolkit[11] with local language support, a localized desktop and office productivity applications, all in Malayalam.[12] The Kerala Bureau for Industrial Promotion (KBIP) undertook the work in association with FSF India. The menu and other texts, like messages, were translated into Malayalam so that a person who knows only Malayalam could comfortably use a computer with the customized GNOME desktop. Unfortunately, this work was never released to the public because of official apathy.

Another Free Software based development was in connection with the Modernizing Government Programme (MGP). MGP was drawn up as part of the strategy of the state government to overhaul and improve its services. One of the components of MGP was monitoring projects funded by the government. Keltron (Kerala State Electronics Development Corporation) developed the Programme Performance Monitoring, System (PPMS) for tracking the performance of various departments as part of MGP. PPMS contain four major projects. The first project, PPMS1, is a performance monitoring system for 17 government departments covering a total of 93 initiatives, 50 in the first phase. The system uses results-based management methods to measure performance based on impacts, outputs, outcomes and activities. PPMS2 is a set of service delivery projects. It addresses performance monitoring of 2,584 institutions state-wide like schools and community health centres, primarily in the areas of fund flow management, administrative payment orders and so forth. The third project is a human resource module named *e-bandham* (e-contact). It monitors attendance, leave, and travel allowances of the programme support executives. The fourth project is *Sevanamudra* (Sign of Service), Quality Improvement Programme and Performance Certification Mechanism for government institutions.

Sutharya Keralam (Transparent Kerala) is another project undertaken using Free Software. This is a Right to Information initiative of the Government of Kerala to ensure transparency and efficiency in everyday functions of the government. The major objectives of the project are the automation of Chief Minister's Grievance Redressal Cell and convergence of all the available forms of Communication so as to guarantee People's Right to Information.[13] The Centre for Development of Imaging Technologies (CDIT), an institute under the Government of Kerala developed the project using only Free Software technologies. Other Free Software projects in the State include: 1) computerization of the offices of milk producers' unions that come under the Kerala State Milk Marketing Federation (Milma), 2) a Management Information System for the Integrated Child Development Services (ICDS) developed by CDAC, 3) a District Collectorate Suite developed by the National Informatics Centre and a computerization project of Calicut University done in house. A study on Free Software projects in public enterprises in Kerala has been done by SPACE and is available online.[14]

Free Software in Business

There are several companies in the State doing business using Free Software. We mentioned River Valley Technologies of Radhakrishnan. *Beta Computers,* an enterprise in Kochi, also engages in business using TeX. Open Source Solutions (Kochi) is a cooperative effort that consists of young programmers who were involved with the Ernakulam Industrial Infrastructure Development Project. The project started working to computerise the Panchayats (Local Self-Government Institutions) in the district. They used only Free Software and computerised a few Panchayats. However, the state-wide programme for the same purpose (Information Kerala Mission), which used only proprietary software, superseded their efforts. The youth involved in the project organized a co-operative society and started doing business with Free Software. They developed software called *Sanghamitra* for cooperative banks. This has been installed in a number of branches. This is also licensed under the GNU General Public License. They have been developing other software and are profitable as a business enterprise. Linuxense at Thiruvananthapuram is another Free Software concern. As their website describes it: "We are a GNU/Linux-based Enterprise providing software solutions of exceptional quality using cutting-edge technologies; creating a GNU/Linux ambiance for our distinguished clients in their demanding work environments."[15] They provide support for Asianet, a major ISP in the State, and their website proudly exhibits an appreciation by an Asianet official on the effectiveness of the antivirus support they have provided.

Swatantra Software Solutions and Services (S2S2) is a small business in Kannur district that has been involved in selling Free Software CDs and systems with GNU/Linux. They also provide assistance to schools for installing computers and networks using GNU/Linux. Sujeevan, who runs the company, actively promotes Free Software and has participated in training sessions for teachers organized by the Teachers' Association and by FSFI. He has helped in installing the customized GNU/Linux distribution for schools in existing networked computer labs. There are several other companies that do business using Free Software along with other platforms, for instance – InApp Technologies. However, only a few of them engage in software development and support exclusively using Free Software.

Free Software Free Society

Free Software Free Society originated as the name of a collection of articles written by Richard Stallman. It was very appropriate that this name was chosen for a conference organised in Thiruvananthapuram by SPACE, FSF India, and others. The conference dealt with the application of the Free Software model to broad questions of governance, digital inclusion, development and culture. The conference was supported by Hipatia (a European NGO), Kerala State IT Mission, Free Software Foundation of India, and the Indian Institute of Information Technology and Management, Kerala (IIITM-K).

The conference had its origins at the World Social Forum held in Mumbai during Jan 16-21, 2004. Owing to Arun's contacts there with people from Hipatia, which also promoted Free Software, a plan developed for representatives from geographically dispersed locations (e.g., India and Latin America) to share their ideas about Free Software. This idea evoked interest in countries as such as Brazil, Venezuela, Italy and India, and the Government of Kerala agreed to extend support for such a conference. The website of the conference explained the vision:

> Located at the intersection of Free Software, Development and Society, the FSFS Conference will examine the application of the Free Software for equitable sharing models for intellectual artifacts, and ultimately for human development. The conference will also address, inter alia, issues such as technology access and the digital divide; legal issues; and experiences of using the Free Software model in fields such as music and literature.

The conference was held in the beautiful campus of the Technopark at Thiruvananthapuram during May 28-31, 2004. Felipe Perez-Marti, eminent economist and ex-Minister of Venezuela, delivered the keynote address. Another important participant was Senator Fiorello Cortiana from Italy. At the end of the conference, a declaration was adopted, now known as the Trivandrum Declaration, calling on the "social and political institutions to eliminate systems that hinder the development of the knowledge society".[16]

Richard Stallman has now visited India several times and given lectures in several states, but Kerala is the State he has visited most often, owing to the support in this State for his ideology. The first time Stallman visited Kerala was for the Freedom First Conference in 2001. His next visit was in 2004. He was in India in connection with the World Social Forum in Mumbai and came to Kerala (Thrissur and Ernakulam). At the Cochin University for Science and Technology, he released the *Rachana Unicode* font under the GNU General Public Licence (GPL). At Thiruvananthapuram, he delivered the first EMS Memorial Lecture constituted by Kerala University. At the University Senate Hall on January 24, 2004, Stallman spoke about the danger of software patents. The same day he spoke about copyright law and freedom in science at Centre for Earth Science Studies. On Independence Day, he interacted with the students of the Indian Institute of Information Technology and Management Kerala (IIITMK). During this visit, Stallman met the leader of the Opposition (now Chief Minister) V.S. Achuthanandan, who has been a strong supporter of the Free Software movement. Stallman's latest visit to Kerala (23 August 2006) was a seminar by SPACE in association with Kerala State IT Mission on Free Software for Kerala Development, where he gave the keynote address and had lengthy discussions with Chief Minister Achuthanandan. A report on Free Software Projects in Public Enterprises in Kerala, prepared by SPACE, was released at the function.

Conclusion

Free Software has come to stay in Kerala. The natural question is, 'Why Kerala'? There is no other State in the country where Free Software has made so much of an impact as Kerala. Kerala occupies a special place in the country because of the very different development path that the State has followed. As elaborated in the introduction of the book, Kerala is different from India as a whole in many ways. Wages are much higher than in the neighbouring states. Almost every other family has someone working abroad or in the IT industry in one of the major metropolises. "It is, in other words, weird – like one of those places where the starship enterprise might land that superficially resembles earth but is slightly off", wrote Bill McKibben.[17] Yet at the same time, alcoholism, the suicide rate, and drug abuse are among the highest in India.

Kerala has a history of social reform movements. One of the most prominent movements is that led by Sri Narayana Guru for the upliftment of the Ezhava community, a backward caste group in the caste-ridden social hierarchy of the State. Members of this community were barred from entering Hindu temples and even from studying Sanskrit and the scriptures. He led a successful struggle against these and even established a temple himself. Ayyan Kali led a struggle against oppression of lower castes by upper caste people and the State. Mannath Padmanabhan led a movement by the middle level Nair community and established the Nair Service Society. The Christian missionaries who have been active in the State for several decades enabled even the most downtrodden to receive basic education. The Communist Party helped to liberate workers from virtual slavery and to bring about universal education. The Kerala Sasthra Sahitya Parishad (KSSP) promoted scientific thinking among children and adults and also spearheaded the total literacy movement. Yet it is difficult to understand why such things happened in Kerala but not in other States.

The unique history of this land has helped create a distinctive sense of democracy, equity and social justice among the people in the State. This is evidenced by the occasional violent reactions to events that are perceived as violation of basic rights. Police action against tribals who had occupied government land in protest against the government's inaction in providing them land as promised, and suicide of a student who could not continue her education due to inability to pay fees, are two examples of events that led to major protests. Freedom is a concept close to their hearts and the sense of personal dignity is high. People thus find it easy to perceive Free Software as a fight against exploitation by large software companies. Moreover, the penetration of communication networks (telephone, mobile, Internet) is one of the highest in the country, and two of the newspapers with the highest circulation in the country are in the local language. Thus, there is a high degree of awareness of global events. In spite of the small population of the State, it has produced some of the outstanding writers, cinematographers, cartoonists and journalists in the country.

It is interesting that in the 1970s an eccentric film maker, John Abraham, considered by many as the only genius in Malayalam cinema, produced a film *Amma Ariyaan* (literally translated as *For the knowledge of the mother*), by collecting small donations and exhibited it everywhere free of charge. Like Knuth's TeX, this could be considered as a forerunner of Free Software, since the ideology of Free Software is now extended to creativity in other areas through movements like Creative Commons. It is no coincidence that the Free Software movement flourished in Kerala.[18]

Notes

1. We capitalize the F and S in Free Software to distinguish it from proprietary software that is distributed free of cost, sometimes called freeware. We shall also use the shortened form, FS.
2. More details about Free Software can be found. There is plenty of material at the FSF website.
3. The story is biased by my experiences and limited by my knowledge. Individuals or organizations who have contributed to the Free Software movement in the State may have been left out of this narration, though this is written after speaking to several people who were involved from the beginning of the movement. I apologize for any such omission and assure the readers that they are inadvertent and not deliberate.
4. LaTeX is a set of macros for TeX now commonly used for typesetting, instead of plain TeX.
5. For more details: http: //sarovar.org.
6. See InApp Technologies (http: //www.inapp.com).

7. Emacs is the editor developed by Stallman that is very popular among users of Unix-like systems including GNU/Linux.
8. http://www.linuxtoday.com/newsstory.php3?ltsn=2001-07-19-010-20-PR-CY
9. Chapter 6 provides a detailed description of the use of Free Software in IT@School Project: editors
10. GNOME is one of the various desktops available in GNU/Linux.
11. Toolkits are basic building blocks in creating Graphical User Interface based applications.
12. http: //www.apdip.net/projects/malayalam/
13. http: //www.sutharya.kerala.gov.in/
14. http: //www.space-kerala.org/downloads/foss.pdf
15. www.linuxense.com
16. See www.gnowledge.org. The full text of the declaration is available at the main page of fsfs.hipatia.net/wiki/index.php/
17. http: //www.ashanet.org/library/articles/kerala.199803.html
18. This document may be copied, distributed or republished in any media subject to the condition that this note is also included. The document may be modified and republished under the same conditions provided a reference is given to the original publication.

5

IT@School and Free Software in Education

The Kerala Model

BIJU PRABHAKAR AND ARUN M.

The technology-driven development that manifests itself in all fields takes the world in a state of dynamic transition and changes the way we live, the way we communicate and the way we work. New technologies seem to have the potential to provide access to a vast volume of information and to handle this information more competently. The recent developments in computer technology and telecommunications have revolutionized the field of education. Multi-media, computer simulation, virtual reality and other teaching tools have enhanced the teaching and learning experience, both conceptually and pedagogically. As a result, there is a boom in IT-based educational products. Students and teachers can interact through the computer networks, Internet, satellite television and so on and successfully take part in various educational programmes. Notwithstanding these developments, often the more privileged classes in society reap the benefits of new technologies, as access is limited. It is in this context that Kerala's efforts in circumventing the digital divide are revealing, as they show the way to the rest of the country. Kerala is the first State to actively promote Free Software in

School Education, illustrating how Free Software can be used to the advantage of the people at large.

The model of IT in education for school children is a unique feat for Kerala. In April 2002, the project 'IT@School' was initiated with a view to remodel the conventional teaching methodologies in classrooms through the use of Information Technology (IT). The project aims at imparting computer education to high school students. Over 1.5 million students have been covered under this project, one of the largest IT Education programmes based in schools. The new instructional technologies have enhanced traditional teaching, and have brought about changes in curriculum, content and pedagogy. The General Education Department, Government of Kerala, launched IT@School Project to integrate information and communication technology (ICT) in the school education system. It is also meant to provide an opportunity to children from the economically weaker sections of society to learn computers and digital technology. The highlight of the project is that ICT is primarily used as a tool rather than content in teaching. At the same time, IT is being taught in high schools as part of the project.

Why did Kerala opt to introduce IT as a subject in the classrooms? To answer this question, we need to engage in a critical appraisal of the earlier technology assisted interventions in the school education system in Kerala. In order to improve the quality of teaching and learning methods, the Government of India as well as several state governments have tried continuously to introduce and integrate technology in the school education system. As part of this attempt at technological interventions radio, tape recorders, television, and computers were introduced in schools, but none has produced desired results.

In the 1960s and 1970s, radio was introduced to assist the teaching and learning process in schools, followed by audio tape recorders (RCCPs). The 1980s saw the introduction of 16 mm projectors with educational films. These ventures failed miserably, mainly because appropriate supplemental software-radio programmes or lessons in audio cassettes dovetailing with the curriculum did not support them. The third attempt was to introduce colour televisions and video cassette players in the 1990s. This programme still continues in some schools, so it was not a total failure. However, it is not effectively used as an educational tool in most classrooms, though several softwares are made available through television channels. These softwares usually did not synchronize with the curriculum. For instance, the SIET telecast of the educational video in *Doordarshan*, (the public television channel) from 10.30 am to 11.00 am every day, clashed with the school timings and so most of the students found it difficult to view. Since it provided no alternatives to the students to watch at a more convenient time, the needs and convenience of the target audience were ignored. Further, students who are exposed to the high quality programmes of international channels like *Discovery, National Geographic* and BBC naturally would lose interest in the comparatively poorer '*swadeshi*' versions. The teachers themselves do not take much interest in showing the material available, however inferior it may be, to enhance learning.

The next major attempt in 1994-95 was to supply computers to schools under the CLASS (Computer Learning and Studies in Schools) project of the Government of India. This programme also did not succeed in achieving the desired objectives. We note that computers found no mention in the textbooks of the early 1990s, except for the mathematics textbooks of eighth and ninth standards. On the basis of a specially designed curriculum, it was in 1996 that the Directorate of Public Instructions (DPI) of the Government of Kerala launched computer training to teachers and subsequently the school curriculum also was redesigned.

The following reasons may be identified for the apparent failure of these attempts of technology interventions in classrooms: (1) the 'need' element was absent, (2) lack of suitable educational content that could be easily assimilated by the school curriculum, (3) the minimal involvement of technologically empowered teachers and (4) inadequate system of maintaining the equipment, given that schools could not raise resources to meet ever increasing recurring expenses. However, the most important reason is that the stakeholders never felt a critical need for technological intervention. Whenever such interventions were made, the results were intangible and not immediately visible.

The continuous failure in the use of technology based educational tools in schools ultimately pointed to the fact that the educational system in Kerala relied only on rote learning of students, which did not require their cognitive abilities and practical knowledge. This type of learning system did not really need any supportive audio-visual aid and educational tool. In such a situation such tools, even if available, are not used much in classrooms. Hence, even in the era of ICT, the 'chalk and talk method' is widely employed and the learning process greatly depends on the eloquence of teachers. The conventional learning climate that prevailed in the educational system in India pointed to the dire need of engaging in serious preparation at the school level prior to the introduction of Information Technology as a tool.

Hence, the first phase of IT@School Project was designed and initiated as a preparatory stage with the triple objectives of equipping the schools, making the teachers computer literate, and generating awareness in the minds of the parents, students and all concerned about the need of using ICT as an educational tool. A series of advocacy campaigns were unleashed to enlighten the school authorities on the need for moving towards IT education first and then IT assisted education later, when the ground has been prepared. In 2000-03, the Government of Kerala implemented the project scheme in three phases. The first phase, for the academic year 2000, sought to initiate teacher training. The second phase that followed in the next academic year was meant to introduce IT as a teaching tool. The final phase aimed at full-scale integration of IT as a teaching and learning tool in 2003.

IT@School Project

In summary, the IT@School Project is an innovative attempt for the intensive and judicious use of information technology in the school education system. The project aimed at IT-enabled teaching and learning at the High School level and it envisioned using IT as an instructional tool and not as a subject. The focus was on a complete and systematic overhaul of the traditional education system, with an emphasis on participation and activity, coupled with constant upgrading of skills and knowledge of the teacher. We emphasize that the project intended to transform the teacher as the pivotal point for technology aided teaching and learning. Hence, conceptually, it sought to empower the teachers and enhance their creativity and ingenuity rather than supplying them with readymade softwares. IT@School Project shifted its focus from IT-enabled learning to IT as a subject in the curriculum owing to the lack of infrastructure for IT-enabled learning, the anxiety of the school community as regards the new educational methodology, and their concern about the possibility of a loss of seriousness in the learning process.

Advocacy campaigns were unleashed to ensure proper interlacing of interest of all the stakeholders, such as the teachers, parents, people's representatives, local self-governments, former students and philanthropists. The project adopted a method of decentralization of decision-making power to the schools for running and raising necessary resources for not only setting up the infrastructure but also for meeting the monthly recurring expenditure. Each school was directed to constitute School IT Advisory Council with people's representatives, PTA, school management and teachers. The Head Master was given only nominal power of meeting daily expenses subject to a maximum of Rs. 1,000 at a time. Other financial power vests with the IT Advisory Council, so as to avoid complications of financial transactions. It came as a great relief for the heads of the institutions who were always haunted by the ghosts of audit system. The involvement of IT Advisory Council made the decision-making process happen at the school level and brought in a democratic and transparent system of decision making process. The resources were raised mainly from three sources, viz., (a) collection of IT tuition fees from non SC/ST students, (b) reimbursement of IT tuition fees of SC/ST students by the government, and (c) income generation operation outside the school hours.

Private agencies were also solicited to assist the schools in meeting the cost of setting up computer labs. Build-Own-Operate-Transfer (BOOT) labs were established in schools. The micro-enterprise mission of Government of Kerala among women – *'Kudumbasree'* had also set up computer labs in 150 schools. Thus, it brought a culture of self-reliance, enterprise and entrepreneurship in the education system through multi-pronged strategies. The results were evident even in the very initial stages to the extent of having around 2,000 computers in different schools in 2001, with the number increasing to over 35,000 by 2005. Schools were permitted to collect fees and generate other resources to purchase more computers and look after their maintenance and upgradation. The involvement of local self -overnments, MPs, MLAs and the IT Advisory Council at the school level ensured great transparency, monitorability and accountability.

In order to create the essential element of the 'need' for IT-enabled learning among teachers, students, parents, and, particularly school managements, information technology was introduced as a curriculum subject. Further, IT practical training was given much more weight than theory in the examinations and in 2005 it was made a compulsory paper in the tenth standard. Having generated an awareness of 'need', the next step was to build the infrastructure and empower teachers. In fact, the unique concept mooted by the project was that ordinary school teachers were to be empowered to transform them as the key implementer, which further ensured general acceptance of the IT@School Project. The syllabus was kept simple and easily comprehensible so that it was accessible to the most disadvantaged students, such as those in the tribal hamlets of Idukki Wild Life Sanctuary.

ViCTERS – The Edusat Initiative

When technology assisted education by IT@School Project was nearing completion, Kerala acquired the support of Indian Space Research Organization (ISRO), Government of India, and developed the promising ViCTERS -Virtual Class Technology on Edusat for Rural Schools. Hence, EDUSAT launch perfectly coincided with the IT@School in Kerala. The State Uplinking Station was established at Gorky Bhavan at Thiruvananthapuram and the studio (set up in association with the Government of Kerala) C-DIT beams signals to the *Edusat*. Satellite Interactive Terminals (SIT), capable of two way video and audio interaction between the faculty and participants, have been set up in 35 centres of all the districts of the State. This is now primarily used for teacher training and departmental conferences.

The Receive Only Terminal (ROT), is a variation of the Direct-To-Home (DTH) technology and brings in signals to television sets in a cost effective solution for a virtual classroom, but is restricted to forward interaction only. ISRO adopted the Malappuram district and so all the government schools in Malappuram, from lower primary to higher secondary, have been provided with ROTs and TVs. The project also used its funds to set up ROTs and TVs in 100 government high schools located in remote locations like Attapady Hills. The ViCTERS, Kerala's complete education channel, commenced experimental transmission from 13th February 2006. In three months 1,006 government high schools, even in the remotest corners of the State, set technology aided education in motion. Soon, this initiative will be scaled up to nearly 13,000 schools in the State.

Politics of Technology Choice

The State Council of Educational Research and Training (SCERT), an autonomous organization recognized by the state government, prepared the syllabus for the IT@School Project. The government had neither framed nor adopted any guidelines for the choice of software. The SCERT syllabus prescribed the software by brand and insisted that school computers should have pre-installed Windows9 as the operating system. Hence, it appeared that the IT@School Project patronized and preferred one software brand to other products, which closed off alternatives and inhibited the possibilities of economizing the costs.[1]

In February 2000, Intel Asia Electronics had started a training programme 'Intel Teach to Future' among teachers in various states of India with a view to promoting Microsoft's proprietary software and paying teachers directly. It was an attempt to enter the schools through trained teachers who acted as links with the student community at large without the participation of the Education Department. Later, the DPI and Intel Asia Electronics signed a Memorandum of Understanding (MoU) to spearhead IT@School Project in Kerala, which ushered in the corporate sector in promoting IT Education in the State. Thus, as part of the IT@School Project, Intel could continue with the same 'Intel Teach to Future' training programme among the teachers, but this time with the financial support of the government. By 2002, Intel imparted training to 300 teachers as master trainers for the project.[2] Later the first IT textbook was prepared based on the proprietary software and 18,000 teachers were trained in Windows.

The IT@School Project insisted that the schools develop necessary infrastructure with the support of local self-government and neighbouring communities. Thus, schools had to bear the burden of building the infrastructure and acquiring personal computers. Schools were compelled to purchase a particular brand of computers and had no real choice, since these were prescribed in the syllabus. At this point some schools began to use unlicensed versions of the software, ignoring issues of piracy.[3] Multinational companies threatened and even conducted raids in some schools.

This situation sparked public debate on different options in the use of softwares, which generated some awareness about the potentials of Free Software and its advantage over proprietary software.

The advocates of Free Software in Kerala raised objections in using proprietary software in the public education system. The problems of the school community in using proprietary software also strengthened the arguments against it. Kerala School Teachers Association (KSTA) in collaboration with Free Software Foundation organized meetings to sensitize teachers and those teachers who were affected by the costs and restrictions on proprietary software soon were attracted to Free Software. As a result, the government was forced to include Free Software products such as Linux and Open Office as optional subjects in the curriculum in 2003. The government also changed the prescribed computer syllabus to be neutral, instead of basing it only on proprietary software. Hence, the IT textbook for 9th standard in 2003 contained both Microsoft and Free Software operating systems and the corresponding application softwares. Nevertheless, ignoring the training and development of Free Software, the IT@School Project continued its training programmes in the Windows platform in view of conducting the IT practical examination for the 10th standard within three years. At this juncture then opposition leader Sri V.S Achuthanandan wrote a letter in support of the use of Free Software to the Chief Minister Sri A.K Antony, and KSTA conducted a study on the IT@School Project favouring Free Software. The study succinctly pointed out that most of the teachers are ignorant of the prevailing norms in the use of proprietary software and revealed that around 60 per cent of the schools used unlicensed versions of software and 43 per cent of the heads of the institutions are unaware of the piracy issues involved in it. The IT@School Directorate also revealed that the project had no mechanism to monitor the use of software and almost all the approved training centres used unlicensed versions for their training. Despite these attempts, the government neglected Free Software and the training and teaching continued in the proprietary software.

In March 2004 the DPI introduced an online examination system for the students of the 9th standard on an experimental basis, with a view to begin examinations for the 10th standard in the following year. Windows 98/2000/XP was used in developing the experimental Softexam and this necessitated installing Windows 98 and MS Office 2000 for several schools which had already shifted to Free Software for their IT education. Led by KSTA, schools protested against this move and finally the government officially accepted the demand to include Free Software in the coming years.

Free Software Education

With the introduction of computer education facilities in large number of schools in Kannur district in 2002, *Swathantra* (Free) Software Solutions developed software solutions based on GNU/Linux and technology for diskless workstations using LTSP (Linux Terminal Server Project) servers and ether-boot. The other GNU software packages installed include open office, multimedia, Internet applications, programming tools and many applets for subjects such as physics, chemistry, maths and so forth. After providing basic training on how to use the GNU/Linux based system, the teachers have made a smooth transition.

In 2004, for the first time, the Department of Education initiated training in Free Software. The training was intended to introduce Linux and Open Office, although the first IT textbook made a direct reference to Free Software in 2002. The academic year 2003-04 witnessed the first IT softexam in Kerala School curriculum. Two schools conducted the softexam in Free Software while all other schools depended on proprietary software. IT@School Project gradually started its efforts towards transforming Windows to GNU/Linux platform. In 2005, with the help of Model Engineering College, Ernakulam, the IT@School Project has completed the first phase of teachers' training in Free Software involving around 200 master trainers. At present, it is also involved in the process of conducting the second phase, consisting of district level training for 6 to 10 days. Nearly 15,000 teachers were trained in GNU/Linux and other application softwares.

In 2004, IT@School Project with the support of Free Software Foundation of India and Kerala State IT Mission developed a customized GNU/Linux version called IT@School GNU/Linux. In 2006 a new version 'IT@School GNU/Linux Light', which will operate in computers with low memory, was introduced. Presently, the project is utilizing this customized version in schools. During the academic year 2005-06, the syllabus for standard 8th was exclusively based on Free Software. In the following academic year this was extended to the 9th standard, with a supplementary IT textbook completely based on Free Software and IT-enabled learning. Nearly half (48.5%) of schools conducted the softexam on Free Software platform in March 2006.

Milestones of IT Education in Kerala

1980	Class Project with the involvement of Central Government
1996	Computer Training for Teachers by DPI
1998	State IT Policy for IT Education
2000	Vision 2010 document envisioned IT-enabled Education; First Government order to start a project on IT-enabled Education
2001	State IT Policy with a special reference to promote Free Software
2000-2002	SCERT and SIET organized Teachers' Training with the help of INTEL
2002	Introduced IT as subject in Schools under IT@School Project based on Microsoft Proprietary software
2002-2003	Public debate on the issue of using Proprietary software and Free Software
2003	IT@School curriculum added Free Software Linux as optional for study
2005-2006	Training on Free Software given to Teachers; Free Software Curriculum implemented for the eighth standard

As the project authorities have realized, the Kerala Free Software Community has the technical expertise to accomplish the professed educational goals and the project anticipates a complete shift to Free Software by the academic year 2007-08. The IT@School Project aims at setting up a Free Software Resource Centre at its Ernakulam regional centre, beginning a training programme for 200 master trainers in ten batches. The State is in need of an effective resource group to support efforts to promote Free Software.

Conclusion

The IT@School Project thus envisaged universal access and comprehensive coverage at the grassroots in order to percolate the concept of Information Technology to the very roots of society, enabling dynamism and motivation. The relevance of IT@School Project comes to the forefront, especially in the context of the eroding quality of education in the State. The State of Kerala has always been a forerunner in India in social indices. Issues of access and retention hampering universalization of elementary and even secondary education have been always been tackled in Kerala. But the quality of education remained far from impressive. In areas of professional courses and university education the quality of output is far from the requisite standards. The roots of the malady should be attacked at the lowest rung, i.e., the schools. IT@School Project is the outcome of constant vigil and conscious policy initiatives for toning up education and making it resilient to the needs of the changing times. It ensures a learner-centred approach marked by a new pedagogy and a new recognition of the need for constant upgradation of skills and knowledge of the teacher.

The concept of 'Softexam' ensured the transparency and accountability of the examination system and enhanced the students' faith in the system. The few detractors to this new system of examination, being tried for the first time in India, could not stop it being introduced in a sensitive school final examination like SSLC. The students, parents and, of course, the teachers wholeheartedly welcomed it and for the last two years, nearly 4.75 lakh students have participated and emerged with laurels.

Notes

1. Total cost of the prescribed operating system and the application software specified in the syllabus came to be approximately an astounding sum of rupees 74,10,00,000 (Rs. seventy four cores and ten lakhs – Rs. 3,500 + Rs. 25,000 = Rs. 28,500 x 10 computers per school x 2600 schools). During the resource development drive they ignored the legal and ethical aspects such as copyright of the software. Due to the poor socio-economic condition of the majority of the schools, they were unable to follow the legal aspects involved in the use of proprietary software
2. This was the first attempt by a corporate to intrude into the education system of Kerala. Subsequently, corporations such as Azim Premji Foundation have entered and are involved on a large scale.
3. Even during the training programmes teachers were not appraised of issues of piracy. The trainers' trainings in fact contained some modules on copyright. But this was meant only to protect the interests of software proprietors. Most of the schools had either a single copy of the licensed version of the software or pirated software purchased from hardware dealers.

II

Development Discourse

6

ICTs for Governance and Development

A Critical Review of Kerala's Recent Experiences

T. T. SREEKUMAR

E-governance, along with such other 'epithetized phenomena' as learning, banking and marketing has gained currency in recent discussions on the Information Society. Woolgar (2002: 3) points out that 'epithetizing' various existing activities and social institutions are notions such as virtual, digital, electronic (or simply 'e'), cyber and tele that 'conjure up a future consequent upon the effects of electronic technologies'. In this paper an attempt is made to address the promises and pitfalls of e-governance on the basis of an analysis of the trajectories of massive e-governance projects launched in India in the State of Kerala.

In the case of ICT-based networked governance, deep crevices in internal organization with the incorporation of different actors have appeared in the contexts of projects like FRIENDS and Akshaya, launched by relevant departments of the Government of Kerala. The person-to-state interface and, to a limited extent, person-to-person relations are affirmed by the technology of the Internet. Network Society is a complex social domain of opposing interest groups and a space where some of the political and ideological conflicts in the larger society are manifested in newer forms (Sreekumar 2003c, 2006). The technology itself gets enmeshed in the relatively autonomous logic of the network society. This partly explains inertia rather than inevitable progress envisaged by concepts informing paradigms of rapid social change consequent on the implementation of e-governance projects (Sreekumar 2003b).

The Advent of E-Governance in India

India was one of the earliest to respond to the possibilities of using ICTs in development administration in the developing world. It may be noted that the Indian State began to design and execute rural development programmes with a relatively visible ICT content in the 1970s, while international attention on the potential of harnessing ICTs for developmental activities is a much newer phenomenon. The early experiments are examples of attempts to use ICTs for improvising development planning, a key area of State action in the pre-liberalization era. The Dharampur Sub-District Infrastructure Planning for Development (1977) is one such early example of an attempt to use computer applications for cost optimization and decision-making. The Karwar Rural Development Information System (1984) was yet another initiative formulated with a focus on reducing delay and curbing corruption through a monitoring programme based on computer applications (Kaul et al., quoted in Bhatnagar 1990: 7).

'Electronic governance' became a key phrase in the Indian State's efforts to provide a facelift to its administrative machinery in late 1990s. It is now geared for 'good governance' as part of the structural adjustment strategies dictated by the World Bank and other international agencies.[1] As noted by Madon,

> In the second phase, the implementation of the national IT Task Force and State Government IT policies symbolized a paradigm shift in e-governance policies towards using IT for a wider range of sectoral applications reaching out to a large number of people in rural as well as urban areas. Moreover, there has been a movement towards a greater input of NGOs and private sector organizations in providing services to the public. These projects have been influenced by the increasing focus of international agencies such as DfID (Department for International Development), G-8, the UNDP and the World Bank under the banner of 'E-Governance for Development (2002: 2).

A Ministry of Information Technology was formed and the Central Government initiated some tentative projects aimed at testing the potential for e-governance. Besides the Central Government, many state governments also responded to the possibilities of improving administrative functions by introducing e-governance at different levels. Specialized agencies have also emerged within government to initiate innovative experiments. In 1999, a national conference was organized in Bangalore attended by 75 senior bureaucrats, including IT secretaries of 32 states and union territories in India.[2] The conference affirmed a resolve to create "one-stop, non-stop, efficient, effective, responsive, transparent citizen governance through the use of information technology" (Katakam 1999: 78). The conference produced a declaration on the intent and content of e-governance programme in India, although it was criticized as bereft of specificity and the plans lacking time frames (Ibid). The declaration emphasized the need for shifting the focus of governance from government centric to citizen centric in the wake of the opportunities opened by information technology for large-scale delivery of quality services. The use of IT was expected to facilitate efficient delivery of government services to citizens and business, "to anyone, anytime, anywhere through a variety of channels at a reasonable cost" (quoted in Ibid: 9). The declaration recognized the need for re-engineering the process of government to achieve synergy with technology. A critical factor identified was the centrality of upgrading skills of the existing workforce, while also underscoring the necessity of industry-community-state partnerships in e-delivery of services. A sound communication infrastructure for ubiquitous access, a conscious effort to harmonize IT with regional requirements and innovative use of IT to prevent possible social exclusion were considered essential components of the e-governance strategy. The role of central government in supporting capacity building efforts of State governments was also underlined.

Many projects were announced at the conference although many of them were eventually shelved or totally abandoned for various reasons.[3] However, since 1999, the proliferation of e-governance projects in India has been phenomenal. Surging numbers of what is identified as e-governance projects indicate that most of the state governments and union territories in India claim to have accepted the need for undertaking e-governance initiatives.

Although initial efforts on the part of the Indian State to link ICTs with development projects were limited to localized projects and district level planning, we can also observe that these soon advanced to centralized projects for connectivity and regulation. There are clear signs of a realization that ICTs, while providing a potential for reassuring the State's commitments to the developmental agenda, open a gateway for strengthening the arms of governance. Nevertheless, most of the E-Governance initiatives have focused on development of infrastructure, capacity building and policy changes as well as participation of the private sector, while showing a growing insensitivity towards integrating civil society into its fold (Sreekumar 2003c). In this paper we will take a closer look at the two state-sponsored initiatives in Kerala, namely, FRIENDS and Akshaya.

The characteristics mentioned are true of some of the initiatives such as FRIENDS and Akshaya launched in Kerala. The former aimed to cater to the everyday needs of a wide section of rural consumers. The project, which sets its objective as social engineering and development through ICTs, has marked a paradigm shift in the way government functionaries relate to the needs of the poor. FRIENDS was conceived as single window scheme. This major ICT project was first launched in a region that is largely tribal and impoverished. Nevertheless, it remains an administered programme with little relationship to civil society. No major social organization has been made partners in its implementation.[4]

A Comparative Frame

For a closer understanding of the process of state-led e-governance programmes in India in general and Kerala in particular, I will compare Kerala and Bihar, two states, that differ widely in terms of the social endowments considered to be useful in cementing the structures of e-governance, but are more or less at the same level of progress in this respect. Kerala is often credited with the formulation of an innovative information technology policy and is known for a powerful repository of educated human power, NRI presence and high connectivity. On the other hand, Bihar lags behind Kerala in all fronts. Nevertheless, it is worthwhile to note that Bihar, one of the BIMARU[5] states in India, has made some significant strides in e-governance, even though these are clearly at a very early stage of formulation and implementation. The state government has initiated the e-governance project using funds provided by the Planning Commission, Government of India. A sum of Rs.10 crore[6] was thus transferred to the state government for the financial year 1999-2000. The project is being executed through National Informatics Centre (NIC, operative in the state since 1990).

Table 1

Status Summary of Computerization Projects in Bihar

Department	*Objectives*	*Status*
Transport	Package for speeding up tax collection, registration, renewal of licenses etc.	Completed at the state level. District level computerization of the department under way.
Treasury	Statewide District Treasury computerization in a phased manner	Sites chosen and equipments provided to the identified districts. Training programmes underway. Proposal of worth Rs. 6.22 crores to computerize 55 district treasuries in Bihar under consideration.
Finance	Computerized budgeting using new techniques	Budget since 1999 are prepared digitally. Budget released in a CD-ROM for the benefit of micro level planning and analysis
GPF	Computerized maintenance of GPF subscriptions, loans, withdrawals and refunds of 3,00,000 government employees.	The software package is designed, developed and implemented in Patna. Proposal under consideration for statewide GPF.
Commercial Tax	Computerized management of tax collection	Software to monitor the fund flow, defaulters' list, reminder letters, cases monitoring etc., has been designed.
Bihar State Electricity Board	Facilitating regular and accurate billing and timely collection of revenue.	No concrete proposals under consideration
Urban Development	Computerization of the Urban Basic Services for the Poor (UBSP) Schemes	NIC involved in generating various MIS and GIS based reports to monitor the life style of slum dwellers
High Court	Computerization of the cause list preparation, AO filing etc.	Patna High Court fully computerized. Statewide Lower Court computerization underway.
Human Resource	Computerization of education	Created an MIS of 200000 school teachers including primary and middle schools. NIC has created a comprehensive database on school information, teacher, infra-structure etc. for all the schools of the State.

contd...

contd...

Health	Computerization of administrative and service delivery systems.	Database created for the family welfare and universal immunization programme. Doctors' panel preparation, nurse information etc., for all districts being developed.
Planning	Computerization of state and district level planning activities.	A District Level Monitoring System was designed for facilitating effective monitoring of various schemes at the state level. NIC has developed software for monitoring projects/ schemes at the district levels.
Public Distribution System	Computerization of rationing activities.	Digitized System operative to monitor distribution of food grains, sugar, kerosene, oil etc
State Remote Sensing Agency	Mapping and monitoring of natural resources	Data compiled to facilitate decision making for sustainable development initiatives of the State. In consonance to the DISNIC-NRIS, the formats designed and sent to all the districts for creating Databases.

Source: Compiled from *Computerization Project Status Report of Bihar,* http: //bihar.nic.in/ ITS/ITScenario.htm retrieved on 11 August 2002

Table 1 provides the details regarding the objectives and achievements of computerization initiatives in Bihar. Nevertheless, Bihar is not known for the implementation of any ICT-led developmental project with a strong social content. By comparison, in Kerala it is officially claimed that

> (A)s a part of the government's e-governance initiatives 34 Government Departments with high C2G interface have been identified. Nodal officers have been identified from each department to oversee the computerization activities of the Departments. They have been provided training on implementation of the project. In areas of hardware procurement, software development and training, Total Solution Providers have been identified for each Department and specific guidelines have been issued by the Department of IT to all Departments/ PSUs.[7]

Table 2 will summarizes the actual position of the initiatives. The trajectories of e-governance initiatives in these two states provide interesting scenarios for comparison. One of the most striking aspects of the differential performance is that Bihar has faired better than Kerala in terms of coverage and progress in the computerization of various departments. Bihar also signed a MOU with Government of India much earlier than Kerala. Nonetheless, Kerala initiatives are more balanced in terms of seeking innovative projects with a social content. However, the Kerala experiments in ICT-based social initiatives are far from offering a satisfactory solution to problems that e-governance and e-education are supposed to tackle. The cases of both FRIENDS and IT@School would illustrate this point. While the department-centred approach in Bihar gives less leverage to involve CSOs in its e-governance programme, the socially focused programmes in Kerala have also shown an antipathy to CSOs and their views. In the next section we provide an analysis of the performance of the FRIENDS project in Kerala.

Table 2

Status Summary of Computerization Projects in Kerala

Departments	*Objectives*	*Status*
Treasury	Computerization and Internet connectivity of District and Sub Treasuries	All District Treasuries are online now. A few Sub-Treasuries are also online
Registration	Computerization of land records and provision of citizen services	54 Sub Registrar Offices (SROs) have been identified for computerization. The pilot project has been completed in 4 offices. In 37 out of the remaining 50 SROs, PEARL has been installed*. The computerization of an additional 113 SROs is now contemplated
Motor Vehicles	Computerization and provision of citizen services	The State Government and NIC have signed a MoU for customization of CORE software for implementation of back-end computerization. Regional Transport Office Thiruvananthapuram has been selected for the pilot project for back-end computerization. Software for conducting online testing for Learner's License is ready. They have been installed in a few RTOs on experimental basis.

contd...

contd...

Rural Development	Computerization of Development Blocks	RDNet – the project for networking all the 152 Blocks in the State has been completed.
Local Self-Government	Computerization of local bodies	Information Kerala Mission – The project for networking all the 1,157 local bodies is considered. Software testing and purchase of hardware completed in some panchayats.
Kerala Water Authority	Computerization and provision of citizen services.	The computerization of 13 offices of the Kerala Water Authority and computerization of billing and collection at 9 centres have been completed. Centralized web-enabled billing and collection at Trivandrum is under consideration.
Civil Supplies	Computerized delivery of citizen services.	The project for automation of Ration Cards has been completed. Smart Cards are being issued in 4 Taluk Supply Offices in Trivandrum on a pilot basis.
IT Mission's FRIENDS	Computerized payment utility bills, taxes and fees pertaining to the participating departments.	Functioning in 14 District HQs, Replication at sub district level is considered.
IT Department's Sevana	To provide information on various government schemes, programmes, general information on local bodies, links to important sites and other facts relevant to the rural populace.	State Library Council has launched the first computerized rural information centre at Kallara Grama Panchayat Library in Trivandrum District. 14 Rural Information Centres, one in each district of the State have also been established.

* Package for Effective Administration of Registration Laws

Source: Compiled from http: //www.kerala.gov.in/

The FRIENDS Project in Kerala

FRIENDS is coined as an acronym for Fast, Reliable, Instant, Efficient, Network for Disbursement of Services. Local bodies and seven government departments/agencies such as Kerala State Electricity Board, Kerala Water Authority, BSNL, Revenue Department, Civil Supplies Department, Motor Vehicles Department, and universities have collaborated to form this integrated services centre with a view to 'enable a smooth and transparent C2G interface'. These centres, launched in 14 district headquarters in the State, accept utility bills, taxes and fees pertaining to the participating departments. According to the implementing agencies,

> (I)t is a 'Single Window Scheme' in which the consumer is given the option of paying for the common services rendered to him under a single roof. The major aspect of this concept is to treat the citizen who is paying for the services rendered to him as a valued customer who must be given all the respect that he richly deserves.[8]

As a multiple agency bill collection system, the project envisages facilitating the collection of various payments within a corporation or municipality at a single centre. The major feature of the project is that its software is specifically programmed to accept payments due to different agencies by incorporating the specific rules and regulations regarding remittances pertaining to each agency.

Many of the projects conceived as part of expanding the scope of e-governance in India had a very strong citizen government interface. This predilection has been shaped by a neo-liberal, state-bashing rhetoric focusing on inefficiencies of governmental service delivery mechanisms throughout the developing world. Application of ICTs through innovative organizational experiments was often pointed out as a possible solution to the problem of inefficiency in service delivery mechanisms for various state governments in India. The first experiment of this kind was initiated in the State of Andhra Pradesh called 'e-seva'.[9] Launched in 2001, e-seva was a one-stop-shop for 30 or more government to citizen and business to citizen services.[10] The idea of e-seva originated from the belief that government departments providing these services perform at a low level of efficiency. The added attraction for the concept came from the idea that delivering governmental services like payment of utility bills certificates and various fees and a wide range of similar services under a single roof using computerized billing systems would help reduce conventional delays and thus improve efficiency. Admittedly, the FRIENDS project of Kerala is an emulation of the e-seva model.

The project was launched in June 2000, first in the State capital Thiruvananthapuram, and was later commissioned in the thirteen other districts of the State. The project exhibited some key differences from the way in which administrative and service delivery was carried out in government departments. The most important were functional and behavioural differences that highlighted the unresponsive character of civil service administration in the State, rather than the use of ICTs for service delivery. One such difference was in the working hours. The normal working hours for departments are from 10.00 AM to 5.00 PM, whereas the FRIENDS centres function between 9.00 AM and 7.00 PM. The public will certainly benefit from this arrangement since there have always been complaints that employees in government departments arrive late and leave early.[11] Moreover, the FRIENDS centres are open on Sundays. The officers deputed to FRIENDS centres are re-designated as 'service officers' thereby eliminating the hierarchy typical of government offices.[12] The behavioural aspect was also equally important. Citizens visiting government offices conventionally look upon the civil servants as arrogant and unfriendly. Hence, the employees posted as service officers were given a formal training at the Institute of Management in Government (IMG), Thiruvananthapuram, an autonomous agency sponsored by the State government and specializing in training civil administrators. The curriculum included topics such as personal development and public relations, in addition to the technical skills required for managing the FRIENDS centres.

Another organizational aspect that deserves mention is the decision to engage the members of the women's credit self-help group called *Kudumbasree* in the project. They are engaged to do the menial jobs. Members of the self-help group are women living below the poverty line, using micro financing among members as a strategy of improving their standard of living. The *Kudumbasree* self-help groups are sponsored by the state government through a *Quango* as part of a central government project for poverty alleviation and financially supported by NABARD, a bank specializing in rural credit. The women from the *Kudumbasree* unit associated with FRIENDS are given menial jobs such as "cleaning the office, delivering the receipt scrolls to the offices of participating departments and preparing tea and snacks for officers as well as public" (Madon and Kiran 2002: 4).[13]

Table 3
Activities of the FRIENDS Centres

Department	*Type of Service*	*Number*
Kerala State Electricity Board	Payment of user charge	
Kerala water Authority	Payment of user charge	
Bharatiya Sanchar Nigam Limited	Payment of user charge	
Local bodies	Property tax Professional tax Traders- license fee	
Revenue Department	Building tax Basic tax Revenue recovery	
Civil Supplies	Fee for new ration card and various certificates	
	Fee for trade licenses	11 types
Motor Vehicles	One time vehicle tax	
	Motor vehicle tax	105 types
	Fee for licenses	20 types
	Fee for permits	142 types
	Registration fee	37 types
	Fee for fitness certificate	
University	University examination fee	352 types
	Other fees	96 types

Source: Adapted from Madon and Kiran (2002: 3-4).

The project has been successful in obtaining the partnership of those departments with a high citizen to government interface in their day-to-day activities (see Table 4). IT-enabled payment of fees and bills to these departments under one roof is certainly a very useful functional rearrangement and a citizen friendly project given the fact that the bureaucratic system in India is still thoroughly colonial in character.

However, a quick look at the performance of the centres reveal a picture of stagnation and official negligence. The apathy of the participating departments has been pointed out as the major reason for the inertia. Even sympathetic observers have found the project heading towards failure due to attitudinal problems of bureaucrats.

Vittal (2000: 16) the Central Vigilance Commissioner, after a visit to the FRIENDS centre in Thiruvananthapuram notes that

> The first shocking thing I learnt was that none of the secretaries to the government of these seven organizations had visited FRIENDS! It is saddening and disheartening that the secretaries of the concerned departments are indifferent to the initiatives being taken by the Chief Minister's vision of an IT based Kerala... Indifference perhaps can be tolerated but what was more damaging is destructive negativism. I was told that the secretaries hold the view that the whole experiment must be scrapped.

This observation is validated by the poor financial performance of the 14 centres in the first year of their establishment. Figure 1 provides trends for monthly collection in the FRIENDS centres in 2001. In most of the centres collection levels have stagnated through out the period, while in a few centres it has been extremely erratic. Collection in most cases has stayed within the range of Rs. 10 to 40 lakhs in most districts. In some cases the lack of coordination and obstinacy of the staff has resulted in creating additional burdens for customers. For example, some departments insist that for certificates the charges should be paid in the FRIENDS centres, while previously the facility was available in the same office that gives the

Figure 1

Monthly Collection from FRIENDS District Centred 2001

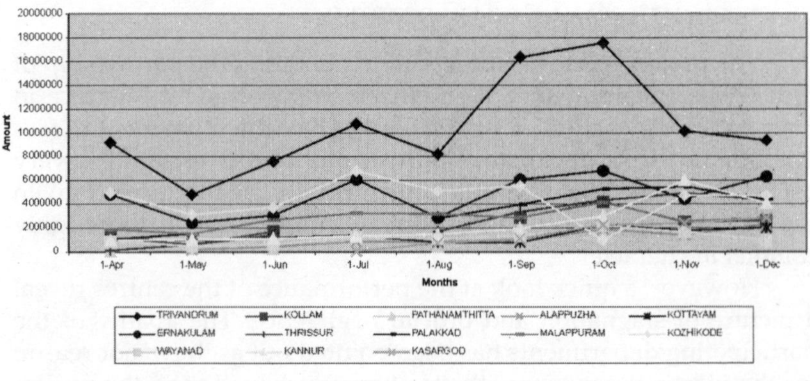

Source: Based on collection figures provided at www.kerala.gov.in (retrieved on 19 October 2002)

certificate. The customer now has to go first to the FRIENDS centre and then visit the concerned department.

Another important aspect of the FRIENDS project is the way in which a *Quango* has been highlighted as a solution to the administrative inefficiencies of government departments. The IT mission that sponsors the project is a *Quango* with personnel, including the Chief Executive, deputed from government departments. It has a minimal and notional link with civil society if one considers the women's unit engaged in carrying out menial jobs as part of it! Moreover, from the errand duties of the women from *Kudumbasree* we find that the scroll rolls are manually carried back to each department and there is no centralized network for the transfer of receipts.

Comparison with SAC, Bahia, Brazil

It will be illuminating at this juncture to attempt a comparison between FRIENDS and the Bahia model in Brazil, a project which can be seen the first ever attempt to run a one-shop centre for citizen services using information technology. The state government of Bahia, Brazil initiated Citizen Assistance Service Centres (SAC) in September,1995, long before e-seva was conceived in Hyderabad.[14] It was an attempt "to bring together federal, state, and municipal agencies in a single location to offer the services that citizens most frequently need and use".[15] The two major attractions of SAC were that it offered citizens time saving schedules to make the payments, while "also delivering services with greater courtesy and professionalism"[16] perhaps not available in government departments. In the case of FRIENDS also we can see that both time saving and courteous delivery of services have been major concerns. But one of the major achievements of SAC was a phenomenal reduction in the overhead expenses of government "since, in many instances, agencies pay much lower rents for space in the SAC than for the properties they previously rented to interact with the public".[17] However in this respect, FRIENDS has not been a success. It is not clear whether any participating departments experienced any reduction in overhead costs as a result of their participation in FRIENDS. These departments have more or less maintained their respective counters and staff for the purpose, indicating more or less the status quo in terms of overhead costs.

In terms of the number of services and the quality of services also SAC far surpasses FRIENDS. It is reported that:

> Now a citizen can register their vehicle or get a driver's license at the SAC. During the same visit, they can get a national identification card, apply for unemployment benefits, look for a new job, get a labor identification card, submit a legal case in small claim's court, get a passport, register a business complaint, check on their retirement eligibility and benefits, etc. Over 500 separate services are offered by the participating agencies.[18]

Twenty-nine different service agencies are part of the SAC system, which include the State Department of Motor Vehicles, the Social Security Ministry, Secretary of Agriculture and Agrarian Reform, Municipal Public Services Secretariat, Labor and Social Action, Public Safety, the Federal Police, Small Claims Court, the State Water and Sanitation Company, and Bahia's private electric company. Another innovative aspect of SAC with no parallel in FRIENDS is a 'Mobile Documents SAC' developed to reach the most remote and deprived communities in Bahia:

> This Mobile SAC is a large, 18-wheel truck equipped with air-conditioning, TV set, toilets, and a covered waiting area. Inside the truck, four basic citizenship services are provided: issuance of birth certificates, identification card, labour identification card, and criminal record verification. When the Mobile SAC arrives in a community, the truck links to the computer network of the SAC headquarters through a telephone line. The truck typically will be parked in the town square, and remains three to four days before moving on the next community on its route.[19]

However, there is a move to initiate FRIENDS centres down to the taluk[20] levels so that coverage of the project can be enhanced. In terms of the question of time saving, a survey has revealed that the average time spent waiting at FRIENDS centres was 5.16 minutes, while the alternative option at the department counters was 218 minutes (Madon and Kiran 2002). Nevertheless, it can be seen that FRIENDS centres lack the coverage and content claimed by SAC.

Akshaya Project

Akshaya Project was officially inaugurated on November 18, 2002 by President of India in the Malappuram district of Kerala with the major objective of spreading computer literacy among the masses. It was based on an entrepreneurship model in which the state government's Information Technology Department would provide assistance to private entrepreneurs to run computer kiosks. Although some early reviews of the project were promising (Antin 2005; Pal and Kiran 2005; Madon 2004), the project had been dogged by several problems from its inception and shows deep cracks in the credibility of official narratives of its success and achievements.

The project began with the characteristic euphoria of all government projects that underestimates the timeline required to complete the different phases. The project could actually begin only six months after the inauguration. When it was finally launched it was believed that total computer literacy could be achieved in the relatively backward district in a span of three months. However it was soon realized that the ground realities are far more formidable than the bureaucrats had initially projected. 'Learners', as the people without computer literacy were designated by the project, were not adequately available for the lessons to start. It took longer than the expected time to find willing participants in the project. Hence many entrepreneurs had to woo people to their kiosks or create artificial lists of learners. Participatory research and orientation work to understand and assess the local requirements, the hallmark of all the major projects in India such as MSSRF Gyandoot or TARAhaat, was not followed systematically if ever in the case of Akshaya. This led to the wastage of resources and time. The first phase, which was expected to be completed in three months, finally took 14-18 months to achieve its stated objective, albeit partially. The campaign mode followed in the initial phase was responsible for the relative success of the e-literacy programme launched by Akshaya. Nevertheless, when the tempo campaign could not be sustained in the second phase, most Akshaya centres found it difficult to get learners for any continued e-education.

The project, as mentioned, had enjoyed some initial success. Entrepreneurs attribute this to the campaign mode and the huge subsidy promised by the project through local governments. But they soon became disillusioned when the local governments failed to pay the dues and they were slowly pushed into a debt trap. They were unable to pay their utility bills and repay loan installments. Many centres had to be closed down. In Thannor panchayat, for example, when I was doing my fieldwork, only four out of the nine centres were functioning. The kiosks that were successful had been functioning before they became Akshaya franchisees and Akshaya was contributing only a minute proportion of their total revenue. It was clear that the kiosks could not be sustained on the basis of Akshaya services alone. In localities where local demand was inadequate for other services or even Akshaya services, kiosks were performing below break-even, which explains the high rate of closures. A media report in March 2005 states:

> Due to the accumulating debts, Akshaya centres are being closed down in Malappuram in large numbers. Only 400 out of the 630 centres opened could survive at the end of the first 18 months. Of these 400, 250 were already in existence before the project started. 50 of them are open but practically with no business. The dues from government to entrepreneurs are in huge arrears. There are entrepreneurs who have to get up to INR 2000 as arrears. One entrepreneur in Ponnani, has to receive an arrear of Rs. 5000 (*Madhayamam Daily*).

One of the major complaints by entrepreneurs, who have now formed an organization of Akshaya franchisees to fight bureaucratic apathy and indifference, was the undue delay in getting the subsidies. The trainees from government departments and learners were expected to purchase books provided by the Akshaya district centre. The entrepreneurs were asked to take bulk orders from Akshaya, which they were unable to sell. As one entrepreneur pointed out:

> We were forced to purchase manuals, which we were unable to sell. I was asked to buy 500 books for my kiosk, of which I could sell only barely 150. Instead of letting us remit the money after selling books they insisted that we should make the bulk purchase from them. This caused huge inconveniences for us and the promise that the balance amount for unsold books will be reimbursed never materialized. I am yet to get my money back (interview with an entrepreneur in Thanoor).

An estimated amount of Rs. 3 million is still due to Akshaya entrepreneurs according some media reports.

IT mission had made promises of collateral free loans from banks to entrepreneurs, which was one major attraction for investment in kiosks. During my fieldwork I found that some entrepreneurs have in fact received collateral free loans while others could not enjoy this benefit. In the case of the former, they have practically stopped the repayment of loan arrears. In the case of the latter they are facing legal action for non-repayment. The office bearers of the Akshaya entrepreneurs are angry and frustrated. One of them said they are unable to launch a massive agitation due to some 'internal differences' but soon will be on the streets if the official apathy continued.

One major government sponsored work that belied the expectations of the entrepreneurs was the computerization of panchayat documents and the health kiosk project. In the case of the latter, it became a non-starter. The projected benefits from such initiatives were huge and helped to create euphoria in the beginning. Although the documentation work with regard to the former has been successfully completed, the entrepreneurs have not received full remuneration. The difficulties in carrying out the work were innumerable, including changes to the software supplied by IKM. All other projects to assist the entrepreneurs including the health kiosk project never materialized. Entrepreneurs also suspect corruption and swindling of public money in the health kiosk project. As one entrepreneur points out:

> I have never seen that machine (the medical equipment for measuring blood pressure). But I have paid for it. They said it was for blood and sugar testing. It was supposed to be kept not in the kiosk but in the panchayat office. I want to get back my money (interview with entrepreneur in Thanoor).

Connectivity is yet another major problem that affects the performance of Akshaya centres. IT mission had opted for broadband connectivity and the work was entrusted with Tulips. 540 centres were to be provided Internet connection for an amount of Rs. 30 million. Nearly 350 centres have been connected to the Internet. But since many centres are closed down Tulips could not complete its work as assigned.

It can be seen that entrepreneurs of Akshaya face several difficulties that require immediate attention by the government if the project is to be sustained. By extending the project to other districts, government has virtually exported the very same empty promises and problems there. The following email exchange in an Internet news group of Free Software enthusiasts [Fsf-Kerala] is a telling example of the popular perception of the project:

Hi,

I am also a candidate to start one centre in my Panchayat. I wanted to get valuable feedback/opinions/suggestions from others. As self-employment for many, it involves huge investment and many are dependent on bank loans(At least I am)!!.Proper help from others will be valuable guidance.

If you are making investment make your backup plans. Don't trust the project.

1. Does anyone (especially anyone from Malappuram) have experience with Akshaya at Malappuram? If so is it possible to share their experience?

I've visited few centres.

2. Did any one use GNU/Linux and Free Software application for setting up Akshaya centres? If so did they face any problem!!!

No. Training programmes and modules from KSITM were in Windows hence almost all of them used Windows. There were some suggestions of having training programmes on Free Software. But I don't think they happened.

3. Does the CDs and other material supplied by concerned authorities work on LTSP thin client machines?

I think they are flash files.

4. Will it demand any requirement for any proprietary software?

You have to talk to officials of Akshaya Project. I think they will allow use of Free Software. If not we will fight for it.

5. What is the current status of the Akshaya Project at Malappuram??.

Best thing you can do is to go to Malappuram and discuss with people there.

Situation is not so good. You will have to make your own business plan. Akshaya will provide you some business. That

> won't break even your project. Don't get carried away by the hype. Do a good study and plan and then start.
>
> *Akshaya Project got Ars Electronic award. I don't know what is the selection criterion. But I am not very much convinced about the claims made about the project in Ars Electronic site.*
>
> (*source:* http://mm.gnu.org.in/pipermail/fsf-friends/2005-July/003426.html)

Conclusions and Discussion

The failure and inertia of state-sponsored projects lead us to the larger question of civil society participation in e-governance projects initiated by Indian state governments, including Kerala and Bihar. Civil society and its representatives have been kept at large in most of these programmes. Even in cases where such participation is sought, the linkages are either weak or notional. In Kerala, in the controversy over the options of choosing Microsoft or Linux-based Free Software for government's IT initiatives, technocrats influential with the government have taken a stand in favour of the software giant. They doubt the expertise and support the State can expect from advocates of Free Software. They even argue that multinational companies led by IBM, Oracle and Sun are orchestrating the controversy. Some experts argue that

> The State should bargain with the companies based on the Microsoft and the Linux after ensuring that there are enough experts in Linux. Advocates of Free Software are silent when it comes to contentious issues like system support. It is important that a system should not collapse midway once it is set up. It is foolish to believe that Free Software users group will solve all your problems free of cost.

While agreeing that the idea of the Free Software Foundation was exciting, the IT department officials try to articulate a pro-corporatist position by claiming, "one cannot be carried away by ideology because security provisions and the propagated ideology would not go together". The contentious nature of the positing cannot be ignored. The controversy has important implications for understanding the state government's perspectives regarding its relations with civil society in ICT projects. It represents a rupture in trust and confidence as well.

While the state-led projects in e-governance show a relative reluctance to work with CSOs, they have attempted to follow the

Quango model wherever relevant and possible. *Quangos* like GyanDoot Samithy and Information Kerala Mission form the backbone of the projects launched by several state governments. *Quangos* are defined as organizations that essentially undertake the responsibility of implementing state-sponsored programmes or public policies. They are funded by the state but operate at arms length from the executive and without an immediate hierarchical relation to it (Theil 2001: 5). The formation of *Quangos* is part of the general strategy adopted by the states, informed by the logic of civil society mediation. As pointed out in the Pliatzky Report (quoted in Flinders and Smith 1999: 29) *Quangos* are created when

> the work is more effectively carried out by a single purpose organization rather than by a government department with a wide range of functions; in order to involve people from outside of government in the direction of the organization; in order to place the performance of a function outside the party political arena.

Nevertheless, *Quangos* like IKM and IT Mission are not real substitutes for CSOs and most often degenerate into behemoth bureaucratic entities. Further, this leads to the incorporation of individuals and organizations that are not accountable to any constituency, even notionally, into the governing structures of developmental and e-governance programmes. The emerging e-governance programmes in Kerala have a complex institutional model where sustainability issues are embedded in the policy prerogatives that mould the state's interest in the programme.

Notes

1. For a critique of the good governance programmes of the Indian government and its technocratic dimension, see Joseph, S (2001).
2. Also see www.ciol.com/content/services/egov. The conference was one of the highlights of the Bangalore IT.com 1999 event.
3. One such example is the project announced by IT Secretary, Kerala for computerizing and networking relevant applications for 1214 local bodies that claimed development of software for the payment of welfare pensions and a database which can be used for procuring building licenses, license fees and taxes. The secretary was confident that the system would help people living in rural areas to "find out

what quotas they are entitled to, what schemes are applicable to them, or in the case of land records, where they are located and whom to contact" (Katakam, 999: 79). Five years after the announcement, the project has not yet taken shape!

4. The Waana wired project in Maharashtra is perhaps a significant counter example. It covers 70 contiguous villages, providing information and knowledge to increase efficiency and productivity of the Sugar Cooperatives. It aims to support the Cooperatives with agricultural and medical information besides setting up a constellation of facilitating centres for continued enskilling and education. The national informatics centre (NIC) of Government of India and state's education department collaborated with the Cooperatives to launch a strong network of fibre-optic cables, V-SATs, PCs, and other ICT equipments.
5. Large and poor north-central states are known by the acronym "BIMORU" or "sick states". (Bihar, Madhya Pradesh, Orissa, Rajasthan and Uttar Pradesh). The acronym "BIMORU" sounds similar to the word for sickness in Hindi.
6. 1 crore = 10 million.
7. www.kerala.gov.in
8. Ibid.
9. Andhra Pradesh is known for its techno savvy Chief Minister's IT initiatives. Trying to emulate the success of Bangalore as the Silicon Valley of the east, Andhra Pradesh has taken many steps to develop Hyderabad, the State capital, as a similar IT hub in Asia. For an analysis of Hyderabad's development as a hi-tech city and the ironies and anomalies of its growth and its consequences for city-hinterland relations, see C. Ramachandraiah, V.K. Bawa (2000).
10. E-seva was the improvised version of a project originally launched in 1999 in the twin cities of Hyderabad and Secundarabad in Andhra Pradesh. While launching e-seva it was claimed that the government would simultaneously take steps to improve infrastructure and bandwidth so that it can be expanded to rural areas that largely remains unfulfilled. For a report see http: //www.apnic.net/mailing-lists/s-asia-it/archive/2001/08/msg00047.html (retrieved on 10 February 2004).
11. The public's reluctance to endorse a strike called by government officers in 2001 for withdrawing some of their financial benefits as part of implementing neo-liberal economic policies in the State is a clear manifestation of loss of people's support for civil servants who

have stood by them in many previous occasions when they struck work.

12. Nevertheless, they draw the same salary and benefits they were enjoying in the parent department. They can also retain their seniority.
13. This is a report of a survey of customers, employees and Kudumbasree units in some selected FRIENDS centres. The survey results and provisional analysis of the data in the form of a report is available at the World Bank site: http: //www1.worldbank.org/publicsector/bnpp. It is not clear if it was a World Bank supported Study. The methodology section of the Report provides the name of a private agency called NORMA in Thiruvananthapuram as responsible for the collection of data. The accuracy of the data and quality of the survey are hence not verifiable.
14. The idea for SAC emerged out of the first annual technology fair in Salvador in 1994 hosted by the Government of Bahia, Brazil. Some government services such as issuing identification cards were offered there, enabled by ICT systems and it was reported that the service was perceived to be far more efficient and hence well received by the public. The idea of delivering these services on a regular basis emanated from this success. See for details http://www1.worldbank.org/publicsector/egov/bahiaSAC.htm

15 Ibid.

16. Ibid.
17. Ibid.
18. Ibid.
19. Ibid.
20. Taluk is an administrative unit below the District, which would include clusters of several villages and towns. There are about 65 Taluks in Kerala.

References

Antin, J.D. 2005. "The Case of Culturally Appropriate Kiosks: An Ecological Approach to Technology and Culture", School of Information Management and Systems, University of California at Berkeley (unpublished http: //technotaste.com/research.htm).

Bhatnagar, S. 1990. "Computers in Developing Countries". In Bhatnagar, S and N. Bjorn Anderson. (eds.) *Information Technology in Developing Countries*. Amsterdam: North-Holland.

Flinders, M. V and M. J. Smith. 1999. *Quangos, Accountability and Reform*. Great Britton: Macmillan Press.

Joseph, S. 2001. "Democratic Good Governance: New Agenda for Change". *Economic and Political Weekly*, March. 24: 1011-14.

Katakam, A. 1999. 'Towards E-Governance". Frontline, 16 (25), 78-80.

Madon, S. and G.R. Kiran.2002. *Information Technology for Citizen-Government Interface: A Study of FRIENDS Project in Kerala*. Washington. World Bank.

Madon, S. 2002. "The Development Impact of E-Governance". Paper presented at the International Seminar on ICTs and Indian Development, Bangalore, 9-11 2002, jointly organized by Institute of Social Studies, The Hague and Institute for Human Development, New Delhi.

Madon, S. 2004. "Evaluating the Development Impact of E-Governance Initiatives: An Exploratory Framework". *The Electronic Journal on Systems in Developing Countries*. 20 (5), 1-13. http://www.ejisdc.org

Pal, J. and Kiran, G R. 2005. "E-Literacy and Connectivity for Development in India: The Akshaya Approach. *Community Technology Review*. Spring-Summer. http://www.comtechreview.org/mt/mt-comments.cgi?entry_id=321.

Ramachandraiah, C. and V.K. Bawa. 2000. "Hyderabad in the Changing Political Economy". *Journal of Contemporary Asia*. 30 (4), 562-74.

Sreekumar, T.T. 2003a. "E-Governance and Social Exclusion", Invited Speaker at the 11th International Anti-Corruption Conference, Seoul, South Korea, May 25-28, 2003.

—. 2003b. " ICTs for Social Change in India: New Development Paradigm or Rural E-topia?" 3rd International Convention of Asia Scholars, Singapore, August 19-22, 2003.

—. 2003c. "New Communication Technologies and Social Exclusion: An Analysis on the Basis of Asian Evidence". International Conference on 'Communication for Citizens towards a Contemporary Society' organized by Camera Municipal de Oeiras held in Oeiras, (Lisbon) Portugal, October 9-10 2003.

—. 2006. "ICTs for Rural Poor: Civil Society and Cyber-libertarian Developmentalism in India". In Parayil. G (ed.) *Political Economy and Information Capitalism in India: Digital Divide, Development Divide and Equity*, UK: Palgrave Macmillan. 61-87.

Theil, S. van. 2002. *Quangos: Trends, Causes and Consequences*. Aldershot: Ashgate.

Vittal, N. 2001. "E-Development Enabling Communities to Shape their Future: The Indian Perspective". In Jayakumar, R et al. op. cit, 1-30.

Woolgar, S. (ed.). 2002. *Virtual Society? Technology, Cyberbole, Reality*. Oxford: Oxford University Press.

7

Understanding Concepts of 'Development' and Linkages to Deployment Strategies for ICT for Development

RENEE KURIYAN AND JENNIFER BUSSELL

Opportunities for using information and communication technologies (ICTs) for 'development' or to 'bridge the digital divide' have been promoted by groups such as the United Nations, the G8, NGOs, governments, the World Bank, and multinational corporations[1] (InfoDev 2002; InfoDev 2003 UNCTAD 2004; HP 2005). For example, in November 2005, at the United Nations World Summit on the Information Society, 174 countries adopted the Tunis Commitment, which is an agreement to build 'a people-centred, inclusive and development-oriented Information Society' so that people can create, access, utilize, and share information and knowledge. This is based on the idea that ICTs can help people to achieve their full potential and to attain 'development' (WSIS 2005).

ICTs have been drawn into the 'development' field as potential tools for poverty alleviation and the economic and social development of urban and rural areas in developing countries (Arunachalam 2002; Eggleston et al. 2002; Keniston 2002; Prahalad

2004; WRI 2005). But, the way institutions implementing 'ICT for development' (ICT4D) projects conceptualize 'development' varies. What does the 'D' in ICT4D projects mean, both ideologically and empirically, for project outcomes? Development is a term that has not been problematized within the ICT4D space. Although actors may assume a consensus on how 'development' is defined with ICTs, this paper demonstrates that this is not the case.

Alternative conceptualizations of both what 'development' is and the way it is linked to ICT strategies have implications for the way actors on the ground attempt to utilize ICTs to meet their project goals. If development refers to a certain level of economic production, then ICT4D project goals may be used to make domestic businesses more efficient and sustainable. If, however, development also includes social and political goals, then projects may be designed to use these technologies to improve access to health care, provide educational opportunities, increase citizen access to government services, and improve basic individual access to communications facilities for the most marginalized groups. In this paper, we ask: How do different institutional conceptualizations of 'development' goals influence the process by which ICT4D projects are designed and implemented?

This research focuses on ICT rural kiosk deployments and examines these projects as development processes. The goal of the paper is to highlight different ideological and deployment strategies of ICT rural kiosk projects and their linkages to development goals. Rural kiosks are computer centres that typically have several computers, owned and run through a variety of models and providing a range of services to users at lower cost than privately owned computers (WRI 2005: 1). The dominant model of kiosk implementation is based on franchises with local entrepreneur ownership. In this paper we highlight various approaches to rural kiosks.[2]

Based on our research, we argue that depending on how 'development' is defined by the institutions implementing kiosks, there can be variations in strategies for deployment and eventually development outcomes of the kiosk projects. We examine different philosophies of development and project goals for three ICT4D projects in India, including a nonprofit organization (Datamation Foundation), a private business (N-Logue), and a public-private partnership in Kerala (Akshaya). We place particular emphasis on the Akshaya Project in Kerala to highlight the challenges of implementing kiosk projects under a broad definition of development, one that emphasizes both social and economic goals. We conclude that there are important social and economic tradeoffs to consider in the achievement of 'development' goals and outcomes, depending on how development is conceptualized and incorporated into ICT4D strategies. The trade-offs between economic and social goals are an important consideration for both policy makers and researchers in their design and evaluation of ICT4D initiatives.

Understanding 'Development' and 'ICT4D'

The definition of 'development' (Lal 1985; Sachs 1992; Ferguson 1994; Escobar 1995; Sen 1999; Mann 2003) has been, and continues to be, widely debated. The most common indicator of development is income per person (Mann 2003: 67). Development in this sense is in terms of an economic standard and could be described more specifically as 'economic development'. The United Nations Millennium Development Goals (MDGs) utilize a comprehensive approach that incorporates poverty and hunger reduction, provision of primary education, gender equality, health issues, and environmental sustainability (UN 2005). These MDGs were adopted at the United Nations' Millennium Summit in 2000 as part of the Millennium Declaration, a document signed by 189 countries with the goals of sustainable development and elimination of poverty. From these goals it is evident that an expanded definition of development includes many social factors including low or no poverty, access to education, access to health care, egalitarianism, and access to a clean and safe environment (Ibid). This conceptualization of 'social development' is much broader than that of its economic counterpart.

Economic and social factors, however, are not the only potential components of development: access to political freedom is also an important consideration (Sen 1999: 5) Contrary to those who emphasize political, social or economic development as a prerequisite for the other, Sen argues that social and political freedoms are "among the constituent components of development" (Ibid.). Thus the broadest conceptualization of development includes economic, social and political development factors in considering whether or not a given country or population is developed and what areas would be viable for development initiatives (see Table 1). These general debates on what development 'is', provide the context for consideration of how to use ICTs to pursue developmental goals. The debate about ICTs and development, however, is a complicated one. While ICTs are seen as potential tools for achieving the MDGs, this is merely one example of how these technologies have been drawn into the development field. Understanding how technology is embedded in social structures, global relations and relations of power are also various lenses by which to understand ICTs within the development context.

Table 1

Perspectives on Development

Economic	*Social*	*Political*
Medium to high per capita income, economic growth	Low poverty, universal education, access to health care and disease reduction, gender equality, environmental sustainability	Guarantees for political and civil liberties, transparency guarantees, protective security

Also relevant to our analysis are critiques of the development discourse itself. Sachs (1992) and Escobar (1994) focus on development discourses and how they shape people's understandings of what they call 'underdevelopment.' These authors are part of the post-development literature that ranges from attempts at Foucauldian analysis of development discourse to reactions against modernity. This literature questions the assumptions underlying development discourse and attacks it as 'self-defeating'. The discourse has been viewed as an imposition of a particular type of knowledge with the idea that this knowledge wields a certain power. Authors argued that the historical production of a discourse created representations of how people come to see themselves as underdeveloped and in 'poverty' (Escobar 1994: 21) and fundamentally contradicted the original goals of development promoters. This emphasis on discourse is important for elucidating how particular understandings and knowledge can be produced and can have material effects. As we will see in the discussion of kiosk cases below, particular understandings of what development is and who should be the focus of development initiatives can have material effects, such as particular ICT project strategies. At the same time, this post-development literature has been critiqued for being ahistoric, holding crude conceptions of power, antipathy towards the State, and for calling on the 'local' as the solution to power struggles (Hart 2001: 650-51). Understanding the various ways development has been portrayed, examined and critiqued is useful to this analysis of ICTs and development, because it highlights that these projects are not simply technical solutions but need to be considered in light of the prodigious scholarship on development, the many ways development has been conceptualized, and the ongoing debates involved.

The specific question of how ICTs may be useful in facilitating development objectives has received significant attention in the last two decades. The World Bank, governments and other institutions promote ICTs as a technological tool to be used to meet the MDGs. Although the World Bank debated in the 1990s whether choosing ICTs drew resources from other development priorities, they now assert that ICTs should not be seen as tradeoffs, but as complementary to development goals (World Bank 2003: 7). Partnerships among private and public actors are increasingly emphasized in the development space as tools for achieving broad developmental goals, such as in the case of ICTs. The Bank measures the success of ICTs in terms of how much progress has been made towards MDGs. ICTs are thought to be useful in 'eradicating poverty' by stimulating macroeconomic growth, making markets more efficient, improving social inclusion, and facilitating political involvement (World Bank 2003: 8).

These goals are not mutually exclusive, but some institutions conceptualizing strategies for ICT4D emphasize particular goals over others. Furthermore, there is a debate as to whether ICTs should be used as part of a holistic and integrated development and growth strategy or as part of targeted approaches to poverty alleviation. On one side, there is the perspective that ICTs are just one means to serve information needs and a holistic and integrated approach is needed to understand the information chain and determine areas for intervention, which requires a series of resources (Duncombe and Heeks 2001: 2). On the other side is the notion that these strategies rely on trickle down effects in poverty and that interventions should instead be focused on the specific needs of local groups (Gerster and Zimmerman 2003: 2). However, these ideas should not necessarily be seen as contradictory. There continues to be much debate and discussion on which approaches are more appropriate given the overall institutional goals of organizations or projects. Thus, it is important to consider particular institutional philosophies and their influence on the design and deployment strategies of ICTs, because of the effects these strategies can have on the basic success of an implementation in terms of who benefits and the financial sustainability of projects.

We will now examine three ICT4D projects to highlight the differences in conceptualization of development and the implications for rural kiosk implementation and deployment. These projects were selected for their varying conceptualizations of 'development' goals. For the Datamation Foundation, 'development' is based largely in the social development characteristics outlined above, but their specific emphasis is on gender empowerment. The promotion of women through the use of ICTs is the key development objective of the Foundation. As a result, their kiosk initiative emphasizes operational characteristics and programme content that are designed to support women's development, even if this means minimal services for men and dependence on subsidies to continue the programme. In contrast, the N-Logue kiosk programme places explicit emphasis on promoting economic development in India's rural areas. Kiosks are owned and operated by local entrepreneurs who aim to provide whatever services will make their kiosk sustainable in the local market. This model involves no subsidies and there is little or no attempt to ensure that particularly disadvantaged groups will have access to the kiosks. The definition of 'development' utilized in the Akshaya Project of Kerala brings together social and economic development goals. In attempting to achieve these goals the Kerala State Government implemented an entrepreneur-based model, but they also subsidized e-literacy training to ensure that all citizens would have the skills needed to make use of the kiosks. In practice there have been challenges for the Kerala Government and kiosk entrepreneurs to find a balance between these developmental goals. This difficulty highlights the importance of recognizing how the kind of 'development' evoked in a project can affect eventual project outcomes.

Strategies for Rural Kiosk Implementation

Datamation Foundation (NGO Model)

The Datamation Foundation was formed by the founder of Datamation Consultants Pvt. to promote gender rights, using technology and rural ICT kiosks in particular, to help and empower women. This project emphasizes one aspect of development as embodied in the MDGs, which is to promote gender equality and empower women. The Foundation also strives to use technology in ways that promote broad social development and poverty alleviation, but the focus is on women's issues.

One of the Datamation Foundation's programmes is the deployment of rural kiosks or Village Information Centres (VICs). The foundation owns the kiosks and employs young women to run the centres. The Foundation determines locations for the centres, based on analysis of the socio-economic and demographic factors. The Foundation tries to put the centres in areas where they will be able to help the most disadvantaged women. The young women chosen to run the centres are also selected based on factors that emphasize empowering the most disadvantaged groups. This reflects the fact that the goal is not to make the centres profitable, but to use the resources of the foundation to provide services to the community. This emphasis on social, rather than economic, goals is an important characteristic of these centres that differentiates them from some other kiosk initiatives. There are currently about 35 kiosks and they do not plan to expand much beyond this number because they do not want to overextend the resources of the organization. Instead the goal is to focus on particular communities and to have a large impact in the few areas in which they are working.

The multipurpose centres incorporate access to computers and the Internet in addition to non-technical classes such as embroidery and sewing. The foundation has also developed a set of digital educational programmes that women can access from computers at the centres. The programmes include business development, such as learning how to make foods that can be sold in markets, and courses on women-specific issues such as reproduction and health. These programmes are mainly available in Hindi and English. There are also illustrated images to help convey the information. A part of the reasoning for business development courses is to enable women to have an income separate from that of their husbands.

The operations of the centres also reflect the goal of promoting women. In some of those centres, particularly those in largely Muslim communities, only women are allowed to use the centres during the day. From 8 AM to 4:30 PM the centre is open for women exclusively, while during the evening hours men are allowed to use the centre. The financial resources of customers are also taken into consideration. While there are charges for many of the services, there is a sliding scale for fees such that the poorest individuals can use the centre for free, those just below the poverty line pay Rs. 50 per month, and those with more income pay more.

By placing an emphasis on women's development goals, the Datamation Foundation has made clear choices in implementation in order to achieve these goals. By limiting access to the centres during the day, the project hopes to overcome any intimidation or social barriers that would prevent women from coming in when men are present. In providing content that is specific to women's interests, the project ensures that there are reasons for women to visit and benefit from the centres. Finally, by charging fees on a sliding scale, the centres enable women without their own income to take advantage of these services. Many other kiosk projects do not offer similar women-friendly environments.

N-Logue (Business Model)

An Indian start-up called N-Logue Communications takes a private sector approach to ICT projects for development and created a for-profit business model to tap into the rural demand in India for connectivity. The Telecommunications and Computer Networking (TeNeT) Group of the Indian Institute of Technology in Chennai incubated N-Logue. N-Logue's business model fits into the Institute's strategy for developing and "disseminating innovative, affordable communication technologies to the rural poor of developing countries" (Prahalad and Hammond 2002: 32). N-Logue's conceptualization of development is based on economic growth, increased income for entrepreneurs, and provision of services for the rural population. The institutional goals of N-Logue are commercially driven, focused on penetration. The selection of entrepreneurs and the services offered are targeted towards financial sustainability.

N-Logue uses a franchise-based business model that has three levels of interdependent networks. First, N-Logue develops and facilitates relationships among a range of organizations to enable and support the business of franchise owners. This includes hardware providers, NGOs, content providers and government. Second, N-Logue works with a regional network of franchised Local Service Providers (LSP), usually recognized business people or district governments. This is a partnership in which the LSP works simultaneously with N-Logue to set up nodes through which kiosk operators can connect. Finally, at the crux of N-Logue's business, the LSP recruits local entrepreneurs to establish village-level kiosk franchises. These entrepreneurs are individuals who can afford to take a loan to begin a business, are educated at least to 12th standard, and demonstrate the ability and motivation to run a centre. The company sells 'kiosk packages' to entrepreneurs through the LSP, which consist of a computer, printer and backup battery. N-Logue provides kiosk owners with training, support, and general technical assistance. However, local entrepreneurs must take responsibility for developing additional product and service offerings as well as marketing strategies. Typically the kiosks provide Internet and telephone access to local populations or villages, in addition to a variety of government and private sector services. These services vary depending on the kiosk owner, but can include computer courses, net-based services, and employment. N-Logue targets the general rural population, but with emphasis on those who can pay for services, not the poorest groups in the community. There are currently N-Logue operations in Tamil Nadu, Karnataka, Andhra Pradesh, Maharashtra, Madhya Pradesh, and Gujarat, and the company continues to expand (N-Logue 2006).

N-Logue emphasizes economic development and the creation of jobs in rural communities. As a result, the entrepreneurship model serves to develop a new source of employment in these villages. However, this employment is only available to certain groups of people because of requirements for the characteristics of entrepreneurs. In addition, services are developed within the centres to drive employment, such as local Business Process Outsourcing (BPO) opportunities for additional people within the community. The fee-for-service model is also intended to support financial sustainability of the kiosks and thereby a viable business for the entrepreneur.

Akshaya (Public/Private Partner Model)

The final project we examine is Akshaya, which was initiated by the Government of Kerala in 2002 as a partnership in collaboration with private entrepreneurs. It is a rural kiosk project to 'bridge the digital divide' to provide access to communication in rural areas as well as to provide 'e-governance' services such as the remote payment of bills and access to official documentation such as birth and death certificates (www.akshaya.net). The goal was to address issues of access to technologies, skills and content as well as penetration of the technologies (IIITB 2005: III. 2; Pal 2005: 3; Kuriyan 2006: 1). In terms of 'development', the Kerala State Information Technology Mission has social development goals of social inclusion, universal access, and education for the 'masses' as well as economic growth through enterprise development and employment.

The State took an active role in initiating the programme and at the same time tried to establish a balance so that local entrepreneurs could own and run the telecentres. The project aims to provide accessibility to ICTs in rural areas and established 630 Akshaya centres through entrepreneurs in the Malappuram district of Kerala. These were supposed to serve approximately 630,000 customers. Originating from a proposal of the local government for 100 per cent district wide e-literacy training, the Department of Information Technology in the Government of Kerala used a private-public partnership model with the main goal of promoting e-governance throughout the State. The State adopted a strategy of computerization of administrative tasks, which could be highly visible and implementable with citizen interface projects. This programme was initiated in this District with the idea that it would be a pilot project that the government would eventually 'roll out' to the 13 other Districts of the State.

In order to achieve both its social and economic development goals, the Akshaya Project incorporates multiple activities to build citizen awareness and develop services to meet local needs. First, the Kerala IT Mission conducted a research phase in which they initiated surveys at a local level and included local government input before launching the project in the Malappuram district of Kerala. This highlights the aspect of doing a full assessment of local needs and understanding local conditions and implementing a strategy that is historically and geographically specific. Second, during this period they created a development strategy that incorporated local needs and attention to marginalized groups and included a plan for economic and social sustainability of the project. They decided to use a model by which entrepreneurs owned and operated the kiosks. However, the government also decided to help minimize entrepreneur risks in the venture by supporting the e-literacy period in which they essentially paid for a large proportion of the loans the entrepreneurs took. Third, rather than just marketing the project to the local communities, the State took an explicit strategy to actually invest in the market by training individual households in e-literacy skills. The project started with an e-literacy phase, subsidized by the State, with the goal of training one member from each household in a basic e-literacy course so households could become familiar with and develop a set of skills with computers. The entrepreneurs selected the 'decision-maker' of each household to attend the computer-training programme. The course was highly subsidized by the government (who paid the entrepreneur for each household trained) and the individual member of the household had to pay a nominal fee. Many of the people who attended the training were women due to the large number of men working abroad in this district. Fourth, the e-literacy training created a large-scale awareness of the project, of computers more generally, and an understanding among the local people of the planned project. Although not everyone in the community who participated in the e-literacy training project now uses the centres, this broad awareness is important to the future use of computers, the roll out of the project to new districts and general awareness of the benefits of computers for the rural population.

The State selected entrepreneurs based on a variety of criteria, including education, business skills, computer skills and gender. The State trained entrepreneurs throughout every phase of the project. Thus, training and capacity building were of paramount importance. During the second phase, the entrepreneurs who owned the centres were supposed to provide local people with services including access to government services through e-governance, computer education courses, communications, and web browsing (IIITB 2005: III. 2; Pal et al. 2005: 3; Kuriyan 2006: 2). The entrepreneurs are supposed to maintain the sustainability of the kiosks and provision of services by charging customers for their services. The project is now being rolled out to seven other districts in the State. The process of expansion is headed by the government, which selects and trains entrepreneurs, initiates the e-literacy phase and provides the connectivity infrastructure (Akshaya website). All of these activities are intended to support the social development of the community and the economic viability of the centres themselves (see Table 2).

Table 2

Rural Kiosk Project Strategies and Concepts of Development

Kiosk Project	*Datamation Foundation*	*N-Logue*	*Akshaya*
Conceptualization of development	MDGs broadly, with emphasis on equality for women and women's empowerment	Economic growth through increased income for entrepreneurs and employment services for rural population	Universal access to ICTs, education for 'masses', economic growth through enterprise development, employment; penetration of technology; get people to use government services
Model of ownership	Foundation owned	Private-business owned	Public-private partnership
Target audience	Women, particularly from lower classes	Rural population – those who can pay for services	Rural population-training for all; those who can pay for services
Services offered	Computer courses, business skills, health education, web browsing, embroidery and sewing	Computer courses, net-based services, and employment	e-governance, computer education courses, communications, and web browsing
Payment for services	Yes, on sliding scale based on income and free for poorest customers	Yes	Yes for services; e-literacy training subsidized for households
Operator selection	Selection by the Foundation based on socio-economic and demographic considerations, with emphasis on helping the most disadvantaged women	Local service providers recruit entrepreneurs with at least 12th standard education and perceived ability and motivation to run a kiosk	Selection by government based on education, business skills, computer skills and some social considerations to women
Financial Sustainability	Based on Foundation and fees for use	Based on entrepreneur	Based on entrepreneur
Model of expansion	No expansion planned	N-Logue selects new districts and recruits LSPs	Government selects new districts

Discussion

These rural kiosk projects drew on different conceptions of development and are implemented with different strategies, ownership models, and project goals. Studies indicate that there are political and social challenges to the implementation of kiosk projects in general and maintaining financial sustainability can be difficult (Toyama et al. 2004: 3; Kuriyan 2006: 10). This paper does not advocate a particular definition of development, ownership model or deployment strategy. We recognize that the implementation and outcomes of these projects will vary and also depend on the political economy and histories of development within individual states in which projects are implemented, overall institutional goals, and existing levels of inequality, education and social differentiation. However, we argue that it is important to problematize 'development' in these projects and explicitly recognize that different goals of institutions and conceptualizations of development will determine the outcomes of ICT4D projects. In particular these cases highlight that not only do different conceptions of development affect the characteristics of project outcomes, but there are also often tradeoffs that are necessary in order to accomplish one kind of development goal instead of another.

Specifically in these cases we see that particular development goals imply tradeoffs in implementation that determine two key project outcomes: (1) who benefits from ICT4D projects, and (2) the commercial viability of kiosks. For example, the Datamation Foundation targets the poorest women of the community in their rural kiosk initiatives. This means it is prioritizing social development and access for the poorest part of the female population to the exclusion of men. Because the Datamation Foundation provides donor funding for the kiosks and subsidizes a sliding scale for kiosk use fees, marginalized groups have increased ability to utilize the centres. At the same time, this means that the centres are not commercially viable in the short run, which is a common goal of other projects that emphasize economic development through rural enterprise with local ownership of kiosks by entrepreneurs.

N-Logue provides an example of this latter goal, as a project that is focused on economic development through commercial viability of kiosks and emphasis on a business model. With this kind of focus, it is more likely that the poorest groups of the population will not benefit from the project because the kiosks serve only the population who can pay for services. But in doing so, there is expected to be a greater chance that these kiosks will be able to achieve financial sustainability without subsidies.

In the final case, the strategy of Akshaya is to try to achieve both benefits for the rural population, including the poor, and illiterate groups through the e-literacy process, as well as commercial viability of kiosks. The project adopts dual goals of social development through increased access to computers for rural people as well as financial viability of the kiosks based on market principles. Research suggests that there is a tension inherent between having these two goals at a macro level (within the State) and a micro level (with entrepreneurs and potential consumers). There are specific trade-offs that result if the State or entrepreneurs emphasize one of these goals over the other (Kuriyan 2006: 10). For example, if the State overemphasizes the financial sustainability goals entrepreneurs may need to cater to the population who can generate profits, and are willing to pay for high priced services rather than the 'masses' who may not be able to pay. On the other hand, if the State overemphasizes the social development aspect of the project and entrepreneurs cater to the development needs of the poor, then the State may have to fund these entrepreneurs on an ongoing basis. Additionally, tensions associated with consumers' perceptions of the state and entrepreneurs also are an important consideration. The perceptions associated with State-led development programmes, that of helping poor people with free and low quality services, can be detrimental to the financial sustainability of the Akshaya kiosks. However, for social development goals, the Akshaya association with the state can be beneficial by creating awareness among rural people that kiosks are places where they can receive useful and low cost computer training. This makes it difficult to run a financially self-sustaining ICT kiosk project that also meets social development goals (Kuriyan 2006: 10). Thus, although partnerships are increasingly prescribed as a solution to development issues, the Akshaya case shows the difficulties of achieving broad developmental goals with differing institutional aims of partners.

Conclusion

The goal of this article was to problematize and analyze 'development' in the implementation of ICT4D initiatives. In each of the cases we discussed, there is a clear definition of 'development' that affects the way these technology access projects are designed and implemented. Each of the resulting deployment strategies holds lessons for how varying characteristics of a telecentre initiative can affect the ability of the initiative to meet its specific developmental goals. These development goals are not mutually exclusive, but it is important to recognize that how development is conceptualized will have concrete effects for beneficiaries and financial viability of rural kiosks.

Important policy and research agendas emerge from this discussion. In terms of research, there is an important empirical question to be addressed: Can there be a common set of evaluation criteria for rural kiosk projects given these varying definitions of development? If researchers make an assumption that a single evaluation model can be applied to projects with diverse goals, then this could limit the relevance of the analysis. As we have seen in these cases, emphasis solely on economic sustainability for evaluation criteria would limit recognition of the social development goals being achieved by the Datamation Foundation's kiosks. Alternatively, a focus purely on access for the most disadvantaged classes might elide successes in local economic growth driven by an entrepreneur.

Similarly, researchers can also draw on this analysis to improve current efforts to understand the outcomes of ICT4D initiative. By problematizing development and interrogating at an empirical level how the diverse understandings of this term have practical implications for project implementation, analysts can better evaluate the causes of particular outcomes within and across projects. For policy makers and practitioners, the analysis presented here shows that it is vital to be explicit about the development goals of particular ICT4B projects. These initiatives must be evaluated based on these terms and with clear recognition of the trade-offs embedded in these choices. In doing so, leaders of projects will be able to communicate better within project teams and with external parties about both the logic of project implementation strategies and the relevance of project outcomes.

There is no single model of development for rural kiosks. This article emphasizes that the imposition of one understanding of development by researchers or practitioners can lead to a lack of critical understanding for the different types of developmental impacts (whether they be social, economic, political) these projects can achieve. For ICTs to have the positive effects on development that so many people desire, it is necessary to improve our comprehension of these goals themselves, and to implement and evaluate projects within this more nuanced understanding.

Notes

1. The authors would like to thank Kentaro Toyama and an anonymous reviewer for their comments. This material is based upon work supported by the National Science Foundation under Grant No. 0326582 and Microsoft Research India.
2. This analysis and discussion is based primarily on interviews conducted by the authors, in addition to review of primary and secondary literature on the three kiosk initiatives. More than 60 open-ended interviews were conducted with actors from 11 kiosk projects throughout India (Projects included: Dhan Foundation, nlogue, Drishtee, Akshaya, Bhoomi, Common Service Centres, Rajiv Internet Villages, SARI, TARAhaat, MS Swaminathan Research Foundation, Datamation Foundation). This included interviews with both governmental and nongovernmental actors and project staff.

References

Arunachalam, S. 2002. "Reaching the unreached: how can we use information and communication technologies to empower the rural poor in the developing world through enhanced access to relevant information?" *Journal of Information Science* 28(6), 513-522.

Duncombe and Heeks. 2001. *Information and Communication Technologies and Small Enterprise Development: Lessons from Botswana*. Institute for Development Policy and Management. January.

Eggleston, K., Jensen, R., Zeckhauser, R. 2002. *Information and Communication Technologies, Markets, and Economic Development.* Cambridge: Centre for International Development, Harvard University.

Escobar, Arturo. 1994. *Encountering Development: The Making and Unmaking of the Third World.* Princeton: Princeton University Press.

Ferguson, James. 1994. *The Anti-Politics Machine: Development, Depoliticization and Bureaucratic Power in Lesotho.* Minneapolis: Regents of University of Minnesota.

Gerster, R. and Zimmerman, S. 2003. "ICTs for Poverty Reduction". *Information for Development Online, I4D.* July-August.

Hart, Gillian. 2001. "Development Critiques in the 1990's: Culs de Sac and Promising Paths." *Progress in Human Geography* 25(4), 649-658

Hewlett Packard. 2005. "Global Citizenship Report". Palo Alto, CA: Hewlett Packard.

IIITB. 2005. "Case of Akshaya – Information and Communication Technologies for Development. A Comparative Analysis of Impacts and Costs from India." IITB, Kiran, GR, Bangalore.

InfoDev. 2002. "Monitoring the Digital Divide." Washington, DC: World Bank.

InfoDev. 2003. "ICTs, Poverty, and Development – Learning from Experience." Washington, DC: World Bank.

Keniston, K. 2002. "Grassroots ICT Projects in India: Some Preliminary Hypotheses." *ASCI Journal of Management.* 31.

Kuriyan, R., Toyama, K. Ray, 2006. "Integrating Social Development and Financial Sustainability: The Social and Political Challenges of Kiosks." Proceedings of International Conference on Information and Communications Technologies and Development: ICTD 2006.

Lal, D. 1985. "The Misconceptions of Development Economics." *Finance and Development* 22(2), 10-13.

Mann, C. 2003. "Information Technologies and International Development: Conceptual Clarity in the Search for Commonality and Diversity." *Information Technologies and International Development* 1(1), 67-79.

Markle, Accenture UNDP. 2001. "Creating a Development Dynamic: Final Report of the Digital Opportunity Initiative." Accenture, Markle Foundation, and the United Nations Development Program.

N-Logue. www.n-logue.com Accessed June 18, 2006.

Pal, J., Nedevschi, S., Patra, R., Brewer, E. 2005. "Multidisciplinary Approach to Shared Access Village Computing Initiatives: The Case of Akshaya." *Global Challenges of eDevelopment.*

Prahalad, C.K. and Allen Hammond. 2002 "What Works: Serving the poor profitably." Markle Foundation and World Resources Institute www.digitaldividend.org/pdf/serving_profitably.pdf Accessed June 18, 2006.

Prahalad, C. K. 2004. *The Fortune at the Bottom of the Pyramid: Eradicating Poverty through Profits*. Delhi: Wharton School Publishing.

Sachs, W. 1992. "Introduction." In W. Sachs (ed.) *The Development Dictionary: A Guide to Knowledge as Power,* Princeton: Princeton University Press.

Sen, A. 1999. *Development as Freedom*. Oxford: Oxford University Press.

Toyama, K, Kiri, K; Maithreyi L, Nileshwar, A., Vedashree, R; MacGregor, R. 2004. "Rural Kiosks in India." Microsoft Research Technical Report. Redmond: Microsoft.

UN. 2005. "Millennium Development Goals Report." New York: United Nations.

UNCTAD. 2004. "eCommerce and Development Report." Geneva: UNCTAD.

8

The Globalization of Science

Is the Internet Changing the Careers of Female Researchers in Kerala?

B. PAIGE MILLER

The purpose of this paper is two-fold. First, it addresses the impact of the Internet on the careers of women scientists in Kerala by examining the changes that occurred between 1994 and 2000 on four dimensions of their careers:[1] 1) travel beyond the university or research institute in which they work; (2) access to information and communication technologies (ICTs); (3) professional ties to both domestic and foreign researchers; and (4) productivity in foreign and national journals. Second, because the Internet is largely expected to globalize the practice of science by creating professional ties between scientists in different areas, I address broader issues of globalization.

Using the four dimensions mentioned above, I address the issue of globalization. On the one hand, the Internet may be globalizing the practice of science for female researchers in Kerala by increasing foreign contacts. On the other, the research careers of scientists may remain stratified, with women orienting themselves locally and men globally. Because different countries and researchers will vary in the extent to which they are "more inward or outward looking," there are likely to be different career outcomes emerging from men and women's experiences with globalization (Schott 1993: 203). My findings suggest that access to personal computers and email has increased dramatically for both men and women. However, contrary to arguments that ICTs will revolutionize the formation of relationships between researchers in geographically distant locations, I find that the impact of the Internet on women's careers is uneven. At the same time that women report fewer global ties and fewer publications in foreign journals over time and compared to men, they are also maintaining domestic contacts and increasing their productivity in national journals. While women may be even more locally oriented over time to their careers in terms of travel and contacts, it appears that they are fairly active members of the local scientific community.

Before elaborating on these results, I provide a review of the literature discussing the globalization process generally, then address the globalization of science specifically, focusing on the relational aspects of the globalization process. Next, I assess the theory of social capital, in the form of collaboration, in relation to beneficial career outcomes in science. I then turn to a discussion of four studies on the nature of science in developing areas. Specifically, I discuss their findings regarding the difficulties women experience establishing and maintaining professional ties in a developing world context. After this, I look directly at the role ICTs may play in alleviating these difficulties for female researchers by providing the opportunity to develop social ties beyond their immediate context, leading to beneficial career outcomes. Finally, I discuss the importance of the local scientific context within the larger processes of globalization.

A Global Science

Globalization very broadly refers to the development of social relations and transactions that extend across geographic boundaries. It is a "transformation of social geography marked by the growth of supra territorial spaces" (Aart Scholte 2000: 8) in which nations, institutions, and individuals interact and communicate (Held et al. 1999). An explication of globalization as either an objective process of interconnectivity, a subjective experience uniquely felt by individuals and groups, or a series of projects designed to shape global policy, is likely to take different forms depending on the aspect of society one is examining. There are two interrelated features that mark the globalization process, whether exploring cultural, political, economic or relational dimensions: (1) globalization is uneven and (2) it is shaped by the development and diffusion of technology – particularly ICTs (Nederveen Pieterse 2004). These two features hold significant implications for the changing nature of science, particularly when considering the stratified experiences of male and female researchers in less developed areas.

The globalization of science, although distinctive in the way it is 'experienced and mapped' (Nederveen Pieterse 2004: 14) by men and women in developed and less developed countries, broadly refers to the potential for increased interconnections among scientists located in geographically dispersed settings (Drori et al. 2003; Shrum 2005). Specifically, the notion of a global or communal science implies that information is widely available to researchers and, more importantly, that there is a high degree of communication between them. One way to examine the extent to which science has become global is through measures of collaboration. Collaboration, or the open 'exchange of ideas, research techniques, methods, and knowledge,' (Basu and Kumar 2000: 381) as well as the co-writing of papers, is a well-studied topic (Frame and Carpenter 1979; Kyvik and Larsen 1994; Duque et al. 2005; Lee and Bozeman 2005).

Accounts from sociological studies of science generally agree that collaboration between researchers has been increasing, creating what some argue is a global network of scientists (Basu and Aggarwal 2001; Drori et al. 2005). Collaborative relationships are largely seen as positive for the careers of scientists owing to the increased costs of knowledge production that make conducting research more difficult (Duque et al. 2005; Lee and Bozeman 2005).[2] Simply stated, by collaborating, researchers share the burden. Collaborations take two general forms: local and foreign. Local contacts are those within the immediate environment in which a researcher works, for instance, contacts between researchers in the same university or research institute. Foreign contacts are those outside of one's own country, in this instance, outside of India. Opportunities to establish foreign collaborative relationships usually present themselves through travel for education, training, or conferences. While each type of contact may be beneficial, theories of social capital imply that foreign contacts may be more beneficial as a means to acquire needed resources, as will be demonstrated in the following section.

Collaboration as Social Capital

Lin (2000) defines social capital as 'the investment and use of resources in social relations with expected returns in the market-place' (19). An actor's location in the social structure determines the resources available to him or her. Resources are embedded within an individual's social network, the group of relationships comprising an individual and his/her contacts (Smith-Lovin and McPherson 1993; Portes 2005). Because some resources are better than others for attaining certain outcomes – increased productivity for instance – the characteristics of an actor's social network will determine how successful he/she will be in attaining instrumental and expressive ends. Networks that are more diverse, i.e. composed of "contacts with multiple spheres of activity," provide an actor with a wider variety of resources "from multiple, nonoverlapping sources" (Smith-Lovin and McPherson 1993: 226).

In the context of science, important resources include grant money, research facilities, and information. Scientists employed in more developed, highly functioning research systems are likely to possess access to a greater variety of these resources. Combined with the fact that core research countries, typically located in more developed areas, are at the forefront of scientific innovation and knowledge production, contact with professionals in these countries may encourage the flow of needed resources. Researchers possessing such contacts, furthermore, may fair better on such career outcomes as productivity, than those researchers without such contacts. Cultural, historical, and economic circumstances ensure that not all individuals, institutions, or countries are integrated into 'global networks' to the same degree (Schott 1993; Drori et al. 2003).

Owing in part to structural constraints, women are often socialized to fulfill the roles and duties of the domestic realm. From a very young age, this situates women to access and use resources that are less beneficial for successful career outcomes by exposing them to ties that are less diverse and more family oriented (Smith-Lovin and McPherson 1993). Where social and religious norms constrain the physical mobility and interaction patterns of women, as in India, this pattern is especially evident. One final point bears mentioning: although differences between men and women's social capital are small at the outset, they accumulate over the life course, creating gender stratification in career outcomes (Smith-Lovin and McPherson 1993).

Uneven Access to Social Capital: Women and Science

Much of the available literature on gender stratification in science focuses on women employed in highly functioning, core institutions. These studies, with few exceptions, examine the impact of marital status, number of children, and collaboration on productivity, rank, and pay (Long 1990; Fox 1999; Xie and Shauman 2003). In contrast to traditional ideas about the relationship between familial responsibilities (marital status and number of children) and career outcomes, the majority of these studies find either a negligible or positive relationship between the two, although the results for number of children are less conclusive. Consistent with the theory of social capital discussed above, women's ability to collaborate and develop social networks with their peers absorbs most of the variation in career outcomes between men and women (Long 1990; Fox 1995; Kyvik and Teigen 1996; Kemelgor and Etzkowitz 2001).

While far fewer studies have been conducted in less developed areas, they tend to corroborate these findings (Chakravarthy 1986; Kumar 2001; Connell and Wood 2002; Amancio 2003; Gupta and Sharma 2002; Campion and Shrum 2004). Gupta and Sharma (2002) and Chakravarthy (1986) suggest that women in India do not advance at a rate equal to their male counterparts due to the cultural environment of Indian society. Gupta and Sharma (2002) argue that the patrifocal social structure of India socializes women to subordinate their own needs and desires to that of their family, specifically their husband's family. Because of this, and in spite of the fact that women in India have made a great deal of progress, they continue to face a number of hardships in their careers, including feelings of isolation from their colleagues. Similarly, Chakravarthy (1986) argues that women do not receive positions of prestige or authority owing both to overt discrimination and the dual role expectations placed on them, in spite of their equal levels of productivity to men. Essentially, these two articles argue that it is the structure of Indian society, in many respects, that makes it difficult for female researchers to fully participate in science.

In his study of women employed in universities and national research institutes in four Indian cities, Kumar (2001) finds that particularistic standards result in their concentration at the lower ranks of their respective institutions, in spite of their equal productivity levels. Kumar also asserts that factors related to familial responsibilities do not account for the differences in rank for this context. Campion and Shrum (2004), using data gathered in Ghana, Kenya, and India find that the differences between male and female researchers on most dimensions of academic and professional life are minimal. They are equally productive and report similar professional activities and domestic responsibilities (in terms of marital status and number of children). A deficit does appear, however, on three dimensions: women are less likely to report a doctoral degree, they are less likely to travel outside of their immediate context for education, training, or professional visits, and they are less productive in international journals. Campion and Shrum argue that the local orientation of women to their careers prevents them from developing professional ties beyond their immediate work environment. Stemming from these findings, they conclude that the career difficulties women face stem primarily from deficits in social capital.

Access to ICTs: A Tool for Women to 'Go Global'?

Theoretically, the restructuring of social relations – more interactions in less time, should be facilitated by the development and diffusion of ICTs (Shrum 2005). This may be particularly true for women. Access to and use of ICTs may provide them with the opportunity to develop their careers by making it easier for them to collaborate with geographically dispersed professionals (De Roy 1997; Huyer and Carr 2002). Palackal et al. (2006) argue that by using ICTs, women may actually be able to circumvent the constraints placed on their physical mobility by the patrifocal Indian culture discussed briefly above. Because ICTs facilitate communication across distances without actually requiring physical mobility, Internet access and email use may remove the costs and constraints associated with other forms of communication (Luukkonen et al. 1992; Palackal et al. 2006). However, as Duque et al. (2005) note, very little empirical evidence exists to support this theoretical claim. Although a great deal of research has investigated the relationship between collaboration and access to ICTs, very little of this work has been done in developing countries.

Much of what is available falls along two extremes. On one end are optimistic accounts of ICT diffusion as a panacea for not only scientific problems, but development troubles in general (Huyer and Carr 2002). On the other are those that argue the diffusion of ICTs to developing countries will follow the path of other development programmes, due to the ethnocentric and male-dominated manner in which they are conceived (Escobar 1995; Ekdahl and Trojer 2002). I recognize the potential benefits of ICTs, but realize that access to these technologies depends on the "existing economic, social and cultural systems" of the specific context one is examining (Arun and Arun 2002: 39). Inequalities present in a given context may prevent women from benefiting from these communication technologies.

Before moving to the results, it should be stated that, although the emphasis has been on the development of non-local social ties for women, this does not mean that local contacts are unimportant. In the face of the inequalities generated or, at the very least, exacerbated by globalization, women and other marginalized groups may define their research identities around local issues and needs (Escobar 2001). Indeed, this has occurred in areas other than science. With the increase in income disparities between developed and developing countries, for instance, there has been a rise in what is termed localization (McMichael 1996; Escobar 2001; Nederveen Pieterse 2004). Localization, in this sense, refers to consciousness building efforts among marginalized groups often through grassroots social movements. In the face of what some term a borderless world, the local and the global coexist, altering local relations and issues but not eliminating their significance (Aart Scholte 2000: 8). Instead, "local happenings are shaped by events occurring many miles away and vice-versa" (Giddens 1990: 65).

In the following section, I provide a summary of the context for this study before outlining the data and methodology. Next, I report the results, comparing differences between women and men and over time. First, I summarize the changes that occurred in the respondent's personal characteristics in terms of marital status and number of children, as well as their possession of human capital. Second, I analyze changes in women's travel experiences and access to personal computers and email. Third, I examine whether or not access to ICTs translates into (a) more social ties regardless of location and (b) global or local career orientations by reporting the respondent's possession of both domestic and foreign contacts. Finally, I examine the respondents' productivity to determine if more contacts correspond to improved career outcomes and if foreign ties correspond to visibility in foreign journals. I end by discussing the implication of these findings for the future of global science studies.

Kerala Context

Compared to other developing areas, computer access and connectivity in India is relatively high. In 2000, the dispersion of personal computers per 1,000 people was 7.2 and the number of Internet users was 15.91. Compare this to the corresponding figures for Ghana (3.8 and 7.84) and Kenya (6.4 and 12.7). India also possesses one of the most advanced and largest research systems in the world and ranks relatively high on measures of scientific output (Gaillard et al. 1997; Basu and Kumar 2000). Within India, I focus on the southwestern State of Kerala. Referred to by some within the development community as a model of development, Kerala is well known for its high level of social progress but low level of economic advancement. Three features of Kerala's development history and current policies are significant for the present analysis: the social patterns, particularly what they indicate for the status of women, the emphasis placed on technological progress by the government of Kerala, and the general resistance of the people to globalizing trends.

The status of women in Kerala is remarkably high, particularly when compared with women elsewhere in the country. Women in the region boast a history of educational involvement, and consequently, report literacy rates comparable to men (Franke and Chasin 1994). The low level of fertility and the high life expectancy of both women and men mirrors that of more developed areas. The population of Kerala is also predominantly female, a phenomenon virtually unheard of in the rest of India (Alexander 2000). The second feature of the Kerala context, significant for this analysis, is the well-developed infrastructure of the State. Kerala has both a high density of telephone lines and extensive communication networks. The government of Kerala emphasizes technological progress. Its 2001 *Information Technology: Policy Documen*t highlights the goal of the government to "establish Kerala as a leading IT destination[...] within the next five years [and] significantly enhance direct and indirect employment creation in the IT sector" through the establishment of ICT industry and the building of infrastructure (GoK 2001: 5; Arun and Arun 2002). Finally, the population of the State is politically active and shows a great deal of resistance to the presence of outside corporations and development institutions in the region. Simply put, the State is resistant to globalizing trends. Even with access to ICTs, researchers in Kerala, regardless of gender, may not be establishing international ties, preferring instead to focus on internal, domestic relationships.

Data and Methods

This paper makes use of data gathered from 392 scientists employed in national research institutes and universities in Kerala over two phases. The first phase of the survey, with 89 respondents, was conducted in 1994 before the widespread diffusion of the Internet. The second, with 303 respondents, was conducted in 2000. Although the 1994 study included respondents employed in NGOs, I exclude these respondents from the present analysis as they were not included in the 2000 sampling frame.[3] Both waves of the survey included a broad range of questions pertaining to educational experiences, professional contacts, and research activities. Every effort was made to interview as many female respondents as possible (33 were interviewed in 1994; 113 were interviewed in 2000).

I used chi-square and t-tests to examine differences between male and female respondents over time on four dimensions of their research careers: 1) their degree of localism, 2) their access to resources, 3) their possession of social capital, and 4) their productivity levels. For those variables measured at the ratio level, independent sample means tests were conducted and the F statistic was used to determine significant differences. To test significant gender differences for nominal or ordinal level variables, I used chi-square tests and the Pearson's chi-square to determine significance.

Localism refers to the amount of time spent in or outside of one's routine work environment. Using four variables, I measure the degree to which male and female respondents are local in terms of their past and present physical locations. Fewer days or years spent outside of the immediate, work environment indicates a local orientation and vice-versa for those spending more days or years outside. Because opportunities to establish collaborative relationships often present themselves through travel for education, training, and/or conferences, this also acts as a rough measure of the opportunities to develop network ties. The first variable measuring localism is dichotomous and asks respondents whether they received any degree from a developed country or not. The second and third variables, measured at the ratio level, ask respondents to report the number of years they spent in a developed country for educational reasons or for general purposes. The final variable asks respondents to report the number of days they spent away from their parent organization (university or research institute) during in the past year.

While men and/or women may appear to be more locally oriented over time, access to a personal computer or email may also eliminate the need for physical travel. As such, my second dimension, access to ICT resources, employs two dichotomous variables measuring whether or not respondents have access to personal computers and email. These are the resources with the greatest potential, in principle, to globalize the practice of science for female researchers. To assess respondent's possession of social capital, I use three ratio variables measuring the total number of professional ties and whether these contacts are a) domestic or b) foreign.[4] More than social capital, these variables act as direct measures of a woman scientist's career orientation, whereas the localism variable, for the reasons mentioned above, is more indirect.

The total number of professional contacts is simply a count of all those individual level ties mentioned by researchers as important for their research careers. Local contacts include any ties to researchers within the state of Kerala, excluding those that are within India but not in Kerala. External ("foreign") contacts, alternatively, include any relationship with researchers outside of Kerala, including those contacts within India. Consistent with the logic of this paper, the more foreign contacts a respondent reports, the more global is their orientation to their career, and the more domestic contacts they report the more local their orientation. Examining contacts in these two contexts enables me to determine to what extent globalization and localization are occurring simultaneously and to what extent there is a trade off between the two in this context.

The fourth dimension, productivity, is commonly used to determine advancement and other rewards in scientific careers. I use two ratio level variables asking respondents to report the number of articles published in foreign and national journals. While I cannot directly link the social capital of respondents to their productivity in a causal model, I do assume that certain types of social capital are beneficial for certain types of publication outcomes.[5] As such, I expect the more foreign and/or local professional ties to correspond to greater publications in each of these venues. Particularly for publication in foreign journals, those individuals corresponding with colleagues in locations outside of their immediate context may find it easier to publish in international journals as their ties are more likely to possess knowledge of the publication process. Publication in national journals, while it may not provide a researcher with international visibility, is nonetheless important for the advancement of knowledge.

Before reporting these results, I examine the personal characteristics of the respondents in terms of their sector of employment, educational attainment, marital status and number of children. Progress in scientific careers typically requires an extensive investment in education. As such, I evaluate the percentage of men and women possessing advanced degrees only. I also report the respondent's number of children and martial status. Although most of the literature on researchers in the United States demonstrates either positive or negligible effects for marital status and number of children on the research careers of women, the social structure of India (as well as personal choice) may make it more difficult for women with husbands and/or children to maintain an active career. These two variables, then, act as indicators of the domestic demands women face.

Table 1

Sample Characteristics on Sector of Employment, Domestic Responsibilities, and Education by Gender and Across Survey Year

	1994			2000		
Variable	*Male*	*Female*	*N*	*Male*	*Female*	*N*
Sector			89			303
%National Research Institutes	60.7	72.7		62.6	23.0***	
%Universities	39.3	27.3		37.4	77.0***	
Background Characteristics						
Age	45.20	40.27***	89	46.39	45.45	303
% Who are Married	96.4	87.9	89	96.8	98.2	303
Number of Children Education	1.75	1.59	87	1.71	1.74	302
% Who Have a PhD	89.3	63.6***	89	77.9	76.1	303
% Who Have a Master's Degree	8.9	36.4***	89	17.4	23.9	303

***p<.01 **p<.05 *p<.10

Those variables proceeded by a % were tested using chi-square

Those variables proceeded by the word 'number' were tested using the F statistic

The majority of both men and women in the first wave of the study were employed in national research institutes. In the second wave of the study, the sector of employment for men remained unchanged, whereas women were more likely to be employed in universities. Table 1 also provides information on the respondent's domestic demands, and possession of human capital. Consistent with previous studies, I find very few differences between male and female researchers in the two phases. In 1994, women were significantly younger (40.27 compared to 45.20), less likely to possess the PhD (63.6 compared to 89.3) and more likely to possess a Master's degree (36.4 against 8.9). They also reported fewer children and they were less likely to be married.

In the 2000 survey men and women are even more similar, creating a reversal in some of the trends noted above. First, differences in men and women's domestic status have declined. Although the gender differences are not statistically significant, in the second phase of the study women are more likely to report marrying, and they report more children. Men report marrying at the same rates in both surveys and actually report a slight decrease in the number of children, making the differences by gender even smaller. However, the most striking difference to emerge between the two phases of the study relates to the respondent's possession of human capital. The gender-gap in educational attainment, present in the first wave of the study, disappears by 2000. Women are equally likely to possess both the PhD and the master's degree.

Results

Table 2 presents the variables measuring the degree of localism in travel and their access to computers and email. In both 1994 and 2000, men and women display similar *decreasing* trends in their travel experiences. For all four variables measuring this dimension women spend fewer days and years traveling, though the only variable to emerge as significantly different in both phases of the survey is the number of days respondents spent away from their parent organization. In 1994, female researchers spent 18.82 days away compared to 32.54 days away for men, and in 2000 they spent 15.51 days away compared to 26.24 days for men. Even with men experiencing a sharper decline on this variable, women continue to spend fewer days away than their male colleagues. The gender gap in number of years spent in a developed country for training is

Table 2

Localism in Travel Experiences and Access to Communication Resources by Gender and Survey year

	1994			*2000*		
Variable	Male	Female	N	Male	Female	N
Travel						
% Who Were Educated in DCs	8.9	6.1	89	6.9	2.7	302
Number of Years Spent in DCs	.66	.51	88	.51	.33	303
Number of Years Spent in DCs for Training	.68	.29*	88	.38	.23	301
Number of Days Spent away from Organization in Past Year	32.54	18.82**	87	26.24	15.51**	291
% Who Reported Access to Communication Resources Personal Computer	67.9	50.0*	88	94.7	72.6***	303
Email	5.5	6.3	87	94.2	72.6***	303

***p<.01 **p<.05 *p<.10

DC refers to developed country

marginally significant in 1994 with women spending 29 years and men 68. By 2000, this gap narrows, and is not statistically significant. While it appears that men and women both are becoming more locally oriented in terms of travel, these findings may be must be interpreted in light of ICT resources available to researchers in Kerala that may have removed the need for travel. In other words, women may be communicating with researchers using new technologies.

By analyzing access to communication resources, I find some support for this latter possibility. Over the two survey phases both men and women experienced dramatic increases in their access to personal computers and email. In the first phase, 50 per cent of women reported access, compared to 67.9 per cent of men, and they were equally likely to have access to email – 6.3 per cent of women reported access, compared to 5.5 per cent of men. In phase two, 72.6 per cent of women reported access to both a personal computer and email, compared to 94.7 per cent and 94.2 per cent of men reporting access to these resources. While men and women both experienced a dramatic increase in access to email and personal computers, small to negligible differences in 1994 become statistically significant ($p<.01$) in 2000. It does not appear that women have experienced the same degree of access as men, as they are less likely to report access to either resource over time. Still, these results do not tell us whether or not women are more or less connected with other outside and/or local researchers, and they do not speak to the issue of whether access translates into a deficit in productivity.

Table 3

Professional Contacts and Productivity in Foreign and Domestic Journals by Gender and Across Survey Year

	1994			*2000*		
Variable	*Male*	*Female*	*N*	*Male*	*Female*	*N*
Professional Contacts						
Number of Professional Contacts	3.87	3.58	86	3.37	3.41	303
Domestic	1.38	2.19**	86	1.51	2.19***	289
Foreign	2.49	1.39**	86	1.86	1.22***	289
Productivity						
Articles in Foreign Journals	3.25	1.28	85	2.98	.92***	303
Articles in National Journals	6.25	3.66	85	4.87	4.95	302

***p<.01 **p<.05 *p<.10

To determine whether women's increased access to communication resources corresponds to a more global or local orientation over time, I examined the total number of professional contacts and their locations. The results in Table 3 are somewhat contradictory, suggesting that the relationship between the Internet and the emergence of a global science is not entirely unidirectional. While male and female respondents possess similar numbers of professional contacts in both phases of the survey, they also report a decrease over time in the total number of professional ties. In 1994, men possessed 3.87 professional ties while women reported 3.58. In 2000, they possessed 3.37 and 3.41, respectively. At the same time that communication technologies diffused throughout Kerala, the total number of ties for male and female researchers was decreasing. Is this trend pervasive, or is it dependent on both the location of the contact and gender of the respondent?

The decrease in total number of professional contacts over time is explained mainly by a marked drop in the number of external ties reported by both men and women. From 1994 to 2000, the number of contacts men possessed dropped from 2.49 in 1994 to 1.86 in 2000. Although the decrease for female respondents was less dramatic (1.39 in 1994 to 1.22 in 2000), they report fewer external contacts than men regardless of survey phase. Domestic contacts, alternatively, remained the same for female researchers over time – 2.19 in both phases – and rose for men – 1.38 to 1.51. It appears then that, regardless of gender, Kerala researchers are becoming more locally oriented from phase one to phase two and less globally oriented. I now investigate the possible consequences of this finding on female researchers' career outcomes relative to their male counterparts.

Interestingly, no significant difference existed between men and women in 1994 in terms of their productivity on the two measures used. While women do publish fewer items in each of the venues, the differences are minor – women averaged 1.28 in foreign outlets and 3.66 in national journals, whereas men published 3.25 and 6.25 respectively. Unsurprisingly, the largest publication area for both male and female researchers is in national journals. In the second phase of the study, the publication trends in national journals, while not significantly different by gender, changed with women publishing more articles in this venue (4.95 compared to 4.87). However, what were initially small differences between men and women in external productivity levels become statistically significant by 2000, with all researchers experiencing a drop, but women reporting a sharper decline (0.92 compared to 2.98).

Discussion

Studies of science in developing areas argue that structural, cultural, and individual factors restrict the physical mobility of female researchers, limiting their interactions with other professionals by socializing them to fulfill domestic roles and duties. The consequence is that women report feelings of isolation and possess fewer ties with professionals outside of their immediate work environment. Without access to diverse ties, the argument continues, women are less likely to develop relationships with those who have important professional resources. In turn, this may act to penalize women in terms of career outcomes such as scientific productivity and possibly even rank and pay within an institution (Kyvik and Larsen 1994; Basu and Aggarwal 2001). ICTs may be the tools needed to reverse these trends. The constraints on women's physical mobility and professional advancement may be circumvented through email and Internet use (Palackal et al. 2006).

By examining the changes occurring in female scientists' careers over two periods in Kerala, I sought to address broader questions regarding the theoretical benefits of the Internet for the research careers of women. I located this discussion within the globalization debate, specifically the argument that the Internet will globalize science by creating contacts with researchers outside of their immediate environment. While these questions cannot be answered in causal terms with these data, it appears that the consequences of using the Internet are not straightforward. Like other empirical studies, I found very few differences between men and women on many of the career dimensions examined, including marital status and family size. More interestingly, I found that women are becoming more similar in terms of their educational attainment, as women in the second phase of this study are achieving advanced degrees at the same rate as men.

That said, the 'social connectivity' prospects of personal computers and email has so far had uneven consequences for women in Kerala. It does not seem that access to communication resources has dramatically altered the international contacts of women scientists. The striking increase in access to personal computers and email reported by women was accompanied by a concurrent rise in localism in terms of travel and a decrease in external contacts. The hypothesis that women spend less time traveling because they are maintaining their non-local contacts through computer-mediated communication, then, appears to be false. With fewer days and years spent traveling, it is less likely that women will be presented with the opportunities to develop contacts with professionals outside of their immediate work environment.

Nor does it appear that these differences in foreign contacts can be attributed to the large gender gap in access to communication resources emerging in 2000, as men's greater access to communication resources does not appear to be globalizing their careers. They too report fewer external contacts in 2000. Over this same period, with access to personal computers and email increasing and external contacts decreasing, women's domestic ties remained the same – and actually rose for men.

Access to ICTs in Kerala is not associated with sweeping increases or decreases in social connections, regardless of gender. The same can be said of productivity. Access to communication resources has not been associated with broad increases or decreases on this career outcome. Instead, I find little support for the notion that increases in external contacts will be associated with increases in publication in foreign journals, as women published even fewer articles in this venue in 2000 than they did in 1994. At the same time, they are publishing more in domestic journals.

While some suggest that the Internet will globalize science, creating more connections between researchers, my results suggest a more complicated picture, one in which scientists in Kerala, regardless of gender, are more connected locally at the same time that they are becoming less globally integrated. With said, it does appear that the globalization of science is stratified by gender in this context. Despite improved communication abilities, female researchers in Kerala are becoming more inward looking in terms of their social ties and their production of knowledge, trends that may make it more difficult for them to advance in their careers by limiting their access to the resources embedded within foreign ties. The findings also suggest that as of yet, women have not used the Internet to circumvent the cultural norms of Indian society.

What this will mean in the long term remains to be seen. Since there is no immediate progression from Internet access to greater connectivity with outside researchers, and finally to increased productivity, we must continue to examine the impact of the Internet in different contexts. In light of the fact that women's access in Kerala is not yet as great as their male counterparts, it is especially important to ensure that inequalities between rich and poor countries, and between men and women are not simply re-created along technological lines. Finally, the intent of this paper is in no way to undermine the importance of local contacts and publications. Instead, our findings suggest that, rather than just asking, is the Internet globalizing science for women, we should also be asking, how women "practice the local in the global" (Escobar 2001: 147). Rather than simply testing the association between global orientation and career outcomes, an approach examining what local and foreign ties mean for female researchers themselves might increase our understanding of the changes taking place in women's careers.

Notes

1. This paper presents results from a series of studies conducted between 1994 and 2002 in Kerala, Kenya, and Ghana funded by the Netherlands Development Assistance Research Council and the U.S. National Science Foundation (International Programme; STS Programme; Programme on Information Technology Research). I would particularly like to thank Meredith Anderson, Ricardo Duque, Antony Palackal, Wesley Shrum, R. Sooryamoorthy, and Marcus Ynalvez, without whom this study would not have been possible.
2. Those studies conducted in more developed countries find a fairly direct relationship between collaboration and productivity levels. Kyvik and Larsen (1994), in their study on the effect of international contact on research productivity for Norwegian scientists, found that traveling abroad for conferences, education or training increased productivity, especially in foreign venues. Less consistent findings are available on researchers in developing areas. Duque et al. (2005), in their study of researchers in the developing world, note a paradoxical relationship between collaboration and productivity. Researchers in high collaborating countries report lower levels of productivity, while researchers in low collaborating countries report higher levels of productivity.

3. Removing NGOs from the analysis, I excluded 12 respondents.
4. We do not explicitly examine the structure of the social networks or the resources provided by their networks. Instead, we examine the mere presence of ties to other researchers based on location. While measuring social capital in this way does not directly translate into collaboration, the presence of a tie in an individual's or group's social network, according to Lin's definition, implies that an actor expects, at some point to activate the resources in that tie. In other words, the presence of the contact indicates potential resources.
5. As Lin (2001) maintains, resources within relationships are accessed and utilized in order to achieve some purposive end. Within science, one such instrumental end is likely to be increased productivity.

References

Aart Scholte, J. 2000. *Globalization: A Critical Introduction*. New York: St. Martin's Press.

Alexander, W.M. 2000. "Normal Kerala within Abnormal India: Reflections on Gender and Sustainability". In Govindan Parayil (ed.). *Kerala: The Development Experience Reflections on Sustainability and Replicability*. New York: Zed Books.

Amancio, L. 2003. "Gender and Science in Portugal," *Portugese Journal of Social Science* 1(3), 185-98.

Arun, S. and T. Arun 2002. "ICTs, Gender and Development: Women in Software Production in Kerala". *Journal of International Development.* 14(1), 39-50.

Basu, A., and R. Aggarwal. 2001. "International Collaboration in Science in India and its Impact on Institutional Performance," *Scientometrics* 52(3), 379-394.

Basu, A., and B.S.V. Kumar 2000. "International Collaboration in Indian Scientific Papers," *Scientometrics*. 48, 381-402.

Campion, P. and W. Shrum 2004. "Gender and Science in Development: Women Scientists in Ghana, Kenya, and India," *Science, Technology, and Human Values* 29(4), 459-485.

Chakravarthy, R. 1986. "Productivity of Indian Women Scientists". *Productivity* 27(3), 259-269.

Connell, R.W., and J. Wood. 2002. "Globalization and Scientific Labour: Patterns in a Life-History Study of Intellectual Workers in the Periphery," *Journal of Sociology* 38(2), 167-90.

De Roy, O. C. 1997. "The African Challenge: Internet, Networking and Connectivity Activities in a Developing Environment," *Third World Quarterly*. 18(5), 883-889.

Drori, G., J. Meyer, F. Ramirez, and E. Schofer. 2003. *Science in the Modern World Polity: Institutionalization and Globalization*. Stanford, CA: Stanford University Press.

Duque R.B., M. Ynalvez, R. Sooryamoorthy, P. Mbatia, D.B.S. Dzorgbo and W. Shrum. 2005. "Collaboration Paradox: Scientific Productivity, the Internet, and Problems of Research in Developing Areas," Social *Studies of Science*. 35(5), 755-785.

Ekdahl, P. and L. Trojer. 2002. "Digital Divide: Catch up for What," *Gender, Technology, and Development* 6(1), 1-20.

Escobar, A. 1995. *Encountering Development: The Making and Unmaking of the Third World*. New Jersey: Princeton University Press.

Escobar, A. 2001. "Culture Sits in Places: Reflections on Globalism and Subaltern Strategies of Localization," *Political Geography*. 20 (2), 139-174.

Fox, M. F. 1995. 'Women and Scientific Careers.' In Sheila Jasanoff, Gerald Markle, James Peterson, and Trevor Pitch (eds.). *The Handbook of Science and Technology Studies*. Newbury Park: Sage.

Fox, M.F. 1999. "Gender, Hierarchy, and Science". In Janet Saltzman Chafetz, (ed.). *Handbook of Sociology of Gender*. New York: Kluwer Academic/Plenum Publishers.

Frame J.D. and M.P. Carpenter 1979. "International Research Collaboration," *Social Studies of Science* 9, 481-497.

Franke, R.W. and B.H. Chasin 1994. *Kerala: Radical Reform as Development in an Indian State*. Food First.

Gaillard, J., V.V. Krishna, R. Waast 1997. *Scientific Communities in the Developing World*. Thousand Oaks, CA: Sage Publications.

Giddens, A. 1990. *The Consequences of Modernity*. Stanford, CA: Stanford University Press.

GOK. 2001. *Information Technology: Policy Documen*t. Thiruvananthapuram: Department of Public Relations, Government of Kerala. http:// www.keralaitmission.org/content/downloads/ ITPolicy.pdf

Gupta, N. and A.K. Sharma. 2002. "Women Academic Scientists in India," *Social Studies of Science* 32(5/6), 901–15.

Held, D., A.G. McGrew, D. Goldblatt, and J. Perraton 1999. *Global Transformations: Politics, Economics, and Culture*. Cambridge: Polity Press.

Huyer, S. and M. Carr. 2002. "Information and Communication Technologies; A Priority for Women," *Gender, Technology, and Development* 6(1), 85-100.

Kemelgor, C. and H. Etzkowitz 2001. "Overcoming Isolation: Women's Dilemmas in American Academic Science," *Minerva* 39(2), 239-257.

Kyvik, S and I. Larsen 1994. "International Contact and Research Performance," *Scientometrics* 29: 161-172.

Kyvik, S. and M. Teigen 1996. "Child Care, Research Collaboration, and Gender Differences in Scientific Productivity," *Science, Technology, and Human Values* 21, 54-71.

Kumar, N. 2001. "Gender Stratification in Science: An Empirical Study in the Indian Setting," *Indian Journal of Gender Studies* 8,(1), 51-67.

Lee, S., and B. Bozeman 2005. "The Impact of Research Collaboration on Scientific Productivity," *Social Studies of Science* 35(5), 673-702.

Lin, N. 2000. "Inequality in Social Capital." *Contemporary Sociology — A Journal of Reviews*, 29 (6), 785-95.

Lin, N. 2001. "Building a Network Theory of Social Capital." In Nan Lin, Karen Cook, and Ronald S. Burt (eds.). *Social Capital Theory and Research*. Aldine De Gruyter: New York.

Lin, N. 2002. *Social Capital: A Theory of Social Structure and Action*. Cambridge: Cambridge University Press.

Long, J. 1990. "The Origins of Sex Differences in Science," *Social Forces* 68(4), 1297-1316.

Luukkonen, T.O. Persson, and G. Sivertsen. 1992. "Understanding Patterns of International Scientific Collaboration," *Science, Technology, and Human Values* 17, 101-126.

McMichael, P. 1996. "Globalization: Myths and Realities". *Rural Sociology* 61(1), 25-55.

Nederveen Pieterse, J. 2004. *Globalization and Culture: Global Melange*. Oxford: Rowman and Littlefield Publishers, Inc.

Palackal, A., M. Anderson, B.P. Miller, and W. Shrum 2006. 'Gender Stratification and E-Science: Can the Internet Circumvent Patrifocality,' In C. M. Hine (ed.). *New Infrastructures for Knowledge Production: Understanding E-Science*. London: Information Science Publishing.

Portes, A. 1998. "Social Capital: Its Origins and Applications in Modern Sociology," *Annual Review of Sociology* 22: 1-24.

Schott, T. 1993. "World Science: Globalization of Institutions and Participation," *Science, Technology, and Human Values* 18(2), 196-208.

Shrum, W. 2005. "Reagency of the Internet, or, How I became a Guest for Science," *Social Studies of Science* 35(5), 723-754.

Smith-Lovin, L. and J. McPherson. 1993. "You Are Who You Know: A Network Approach to Gender." In Paula England (ed.). *Theory on Gender: Feminism on Theory*. New York: Aldine de Gruyter.

Xie, Y. and K.A. Shauman. 2003. *Women in Science: Career Processes and Outcomes*. Cambridge: Harvard University Press.

9

Who Has the Internet Empowered?

Rethinking the Relationship between Women and ICTs in the Developing World

MEREDITH ANDERSON

The relationship between women and technology has long been the subject of feminist deliberations. Initially, western feminists questioned the purported benefits of many domestic and reproductive technologies, depicting these innovations as inherently patriarchal in that they reflect and maintain existing patterns of social relations[1] (Wajcman 1991, 1995, 2004; Morgall 1993). More recently, developing world feminists have joined the debate, turning their attention to the complex web of power relations surrounding tech-heavy western development initiatives. The crux of these arguments posits western technologies as laden with patriarchal and colonialist baggage and, therefore, incompatible with the welfare and interests of developing world women (Acosta-Belen and Bose 1990; Gajjala and Manidipudi 1999; Gajjala 2003). This assertion stands in stark contrast to the western development paradigm, which has long promoted industrialization and modernization as solutions for poverty and social inequality in the developing world.

Currently, information and communication technologies occupy centre stage of this platform, having been identified as a means through which to promote development by facilitating international and local communication, improving access to information and enabling international collaboration (Youngs 2002; Shrum 2005). The diffusion of ICT is viewed as particularly beneficial to the status of developing world women, in that structural changes spurred by ICT adoption are expected to undermine traditional sources of gender inequality (Drucker 2001; Kelkar and Nathan 2002). Such promissory arguments, labeled "elixirs" (Davidson et al. 2002), have failed to convince feminist critics with the depiction of these technologies as a panacea for development woes. Rather, "affliction" arguments questioning the value of ICT development policy, reflect feminist concerns surrounding these initiatives such as their tendency to benefit the privileged (Pal 2003) and their inconsistency with the status and roles of developing world women (Mohanty 2003). Indeed, little evidence exists supporting the premise that ICT will radically revise gender relations in the developing world, leaving many skeptics asking "[W]ho has the Internet empowered? How has this happened? How relevant is this process...?" (Gajjala and Manidipudi 1999: 14).

This paper addresses the relationship between ICT and social change under the structural conditions of patrifocality characteristic of the Indian subcontinent. I examine qualitative interviews with professional scientists in Kerala, India, to assess the particular impact of the Internet on their lives and careers over the past decade. Within this context, I aim to attend to feminist skepticism regarding the efficacy and relevance of ICT for development. First, I provide empirical evidence of the effects of ICT on the lives of developing world women. Second, I delimit the scope of my investigation to this specific group of women in order to account for local level variation in sociocultural context long neglected in western development policy.

I propose that these female scientists have begun to use the Internet to circumvent behavioural restrictions that limit access to social resources previously identified by Campion and Shrum (2004)[2] as vital to their careers. This occurrence, significant in its own right as an indicator of social change, also serves to provide much needed empirical evidence of the impact of ICT on the lives of developing world women. Further, by singling out key benefits of ICT in this context, the present findings pave the way for future research as well as local development policy. I do not suggest the results of this analysis be generalized as applicable to women throughout the developing world. On the contrary, I assert that sensitivity to local context is an essential component of effective development policy.

Although the topic of gender and technology is by no means uncharted territory, this assessment narrows the focus to career change among Indian women scientists. I begin this discussion by reviewing the literature on women, technology and development. Next, I examine the Indian patrifocal context and previous research on its impact on the careers of female scientists. Following an account of the research setting and a brief discussion of my methodology, I review the data, focusing specifically on location of higher education, travel abroad, and international contacts. Finally, I discuss the implications of these results in terms of social change within India and in terms of the wider context of development.

Women, Technology and Development

For decades, the impact of western technologies on the Third World has been a theme within development discourse. Initially, the topic of women and development received little consideration in this literature. Gender roles were assumed to mirror those in western countries, with men the primary wage earners and women confined to reproductive roles (Acosta-Belen and Bose 1990). The only western technological initiative directed at women involved family planning and often utilized unsafe or unproven methods of contraception (Morgall 1993). Not until Boserup (1970) proposed that female-inclusive initiatives would increase the efficacy of development programmes did the productive roles of developing world women receive recognition. This liberal feminist call to "bring women into development" did not fully redress the systematic neglect of women in development policy, for, although it promotes the idea that modernization erodes the power and authority possessed by women in traditional society, it fails to question the capitalist ideals (Park 1993) and western-style gender division of labour promoted by these policies (Mies 1986).

Currently, the status of women remains high on the global development agenda, as does the western focus on technology-based initiatives. Gender equality is one of the eight Millennium Development Goals ratified by the United Nations in 2000[3] (United Nations 2005). In response to this call for improvement in the status of women, the western-dominated Group of Eight (G8) proposes well-designed ICT programmes as the ideal method to undermine inequality and create the social and economic networks necessary for development (DoT Force 2001). Utilizing this line of reasoning, Kelkar and Nathan (2002) predict the adoption of ICT development policies will open up new economic opportunities for women by fundamentally altering traditional distinctions between heavy and light, or male and female, labour.

Despite these optimistic predictions, the debate over the impact of technological development initiatives on the status of developing world women continues to evade resolution. Building on the idea that, as products of a capitalist patriarchy, western technologies function to maintain corresponding social inequalities (Wajcman 1991, 1995, 2004; Morgall 1993), critics of these innovations question their ability to improve the lives of developing world women. According to these arguments, western technological initiatives are "saddled with colonial baggage (Gajjala and Manidipudi 1999)," patterned by patriarchal power relations and capitalist interest in production and accumulation, and ultimately exploit developing world women on both global and local levels (Gajjala 2003). In regard to the alleged benefits of ICT, skepticism persists. Pal (2003) hypothesizes that ICTs disproportionately benefit economically advantaged urbanites, while Patel and Parmentier (2005) argue that these technologies adapt to rather than alter the social structure into which they are introduced.

This line of reasoning, it is important to note, does not build a case against technologies themselves, but focuses on the policy-based shortcomings in their role as development initiatives. In this sense, western technologies are not intrinsically harmful to developing world women. As Wajcman (2004) asserts, the relationship between women and technology is neither uniform nor deterministic. Technologies are often adopted for uses not foreseen by their developers[4] and social action has proven effective in shifting the cultural significance of some technologies.[5] Accordingly, Gajjala and Manidipudi (1999) argue that, once removed from their native contexts, technologies are often incapable of functioning in the intended manner and fail to aid development initiatives or empower women. This discrepancy between the theory and practice of technological initiatives is indicative of the ethnocentric nature of western development discourse, which, Escobar claims (1995), fails to recognize that gender stratification in the developing world differs in form and content from that found in western countries. These accounts, Mohanty (2003) contends, also conceptualize gender stratification within the developing world as a homogenous variable.

Ultimately, western development efforts take the form of what Harding (1995) calls "just add women and stir" policies. Lacking sensitivity to local context, these initiatives are expected to produce uniform results despite their inability to properly account for the lives and needs of the women they aim to benefit. If western technologies are ever to improve the lives of developing world women, they must be applied in a manner consistent with the sociocultural milieu. Escobar and Harcourt (2005) suggest that this task can be accomplished by constructing initiatives in accordance with local understanding of concepts such as gender, empowerment, and patriarchy. This focus on cultural variation, part of the larger discourse of the politics of place, "...allows us to talk about women's forms of place-based politics without reducing them to any overarching global logic, but seeing them in their own right and in terms of the translocal effects they may generate through networking and shifts of scale" (2005:6).

The particular utility of the politics of place framework to the present study is the fluid manner in which it conceptualizes power relations (Harcourt, 2005). Within this context, women are recognized as agents of social change capable of challenging and transforming sources of gender oppression through everyday practices of resistance. In this analysis, I address the role of ICT in this process of resistance and social change. Specifically, I focus on the manner in which female scientists in Kerala, India, have begun to use these technologies to access resources previously beyond their reach and, therefore, implicitly challenge the social and religious prescriptions that limit their mobility. By no means should this focus be interpreted as affirmation that ICTs are only capable of improving the lives of the relatively advantaged. Rather, it is likely that the social gains stemming from these technologies will first be evidenced within these groups rather than those with low rates of Internet adoption and use. In the United States, scientists and engineers played an integral part in the early development and diffusion of the Internet (Abbate 2000). Communities of knowledge workers may well assume a similar role in this process in the developing world. In the next section, I provide an account of the patrifocal social order characteristic of the Indian subcontinent. Following this examination of gender stratification in the Indian context, I review previous research on the effects of these restrictions on the careers of Indian female scientists.

Patrifocal Social Order and Scientific Attainment

Sex differences in scientific attainment are not restricted to India. Female scientists throughout the world have long been found disadvantaged relative to their male colleagues in terms of number of publications, access to resources and rewards, and occupational status attainment (Cole and Zuckerman 1984; McElrath 1992; Fox and Long 1995; Fox 1999,1995; Kyvik and Teigen 1996; Xie and Shauman 1998; Ranson 2003). Western feminist assessments of gender and science maintain that the masculine norms upon which early scientific theory was constructed still function to exclude women from scientific careers as well as limit their contributions to scientific knowledge (Fox 1995, 1999; Fox and Long 1995).[6] Lingering perceptions of women as poorly suited for scientific pursuits, Keller (1995) explains, are merely the legacy of the discursive construction of masculinity and scientific objectivity triumphing over femininity and nature. As Subrahmanyan (1998) notes, this explanation does not hold in the Indian context, based as it is on western experience and social order. Gender stratification is the overarching cause of this female disadvantage, but the power relationships underlying sexual inequality in western countries differ drastically from those found in India.

Unlike their western counterparts, Indian women are not limited professionally by cultural perceptions of scientific disciplines as unfeminine (Subrahmanyan 1998). The career obstacles they face are associated with limitations on physical and social mobility (Campion and Shrum 2004). While female competency in mathematics and science-related fields is not questioned, the appropriateness of stepping outside of the traditional feminine gender role and pursuing a career remains contested terrain. According to Abraham (2000), these restrictions reflect the predominant ideology of family orientation, where the collective interests of the extended family receive precedence over the wants and needs of the individual. Within this context, the prevailing image of femininity, which reifies fertility and subordinate status to males, crosscuts caste, class, and religion to define women in regard to male family members.

This distinct brand of male-female power relations constitutes the basis of the larger patrifocal social order found in India. In this system, and distinct from western patriarchy,[7] on the basis of their relationships with men, women can also exercise authority over other women.[8] Patrifocality has been characterized by Gupta and Sharma as:

> gender-differentiated family roles with females being associated with the 'private' sphere; gender differentiated family authority structure (with authority of same-generational males over socially equivalent females, such as husbands over wives, brothers over sisters); family control of marriage arrangements; patrilineal descent, inheritance, and succession; patrifocal residence, with daughters shifting allegiance to husband's family after marriage; and an ideology of 'appropriate' female behaviour that emphasizes chastity, obedience, and modesty (2002: 902).

Patrifocality is not reducible to a fixed set of criteria but fluctuates in content and severity according to a myriad of social factors including region, caste, and class. Central to this ideology is a principal concern with female sexual purity and family responsibility (Chanana 2001; Derne 1994), generally manifested in social restrictions on mobility. Traditionally, Indian women have been prohibited from engaging in extended contact with unrelated men, traveling without an appropriate escort, or residing away from their families (Chakravarthy 1986; Mukhopadhyay and Seymour 1994; Subrahmanyan 1998). According to some accounts, women, especially those with higher levels of education, have begun to gain access to increasing degrees of social freedom such as influence in family planning decisions (Riessman 2000), financial decisions (Shukla 1987), and the gender division of household labour (Ramu 1987; Rout et al. 1999).

Although these recent gains reveal a changing social climate, patrifocal limitations on mobility are by no means a relic of India's past and continue to curb the professional attainment of even the most highly educated women (Campion and Shrum 2004). As Gupta and Sharma's (2002) assessment of gender inequality in India's top scientific institutions illustrates, the discrimination faced by female scientists is inextricably tied to the sociocultural context in which they live and work. Many of the female faculty they surveyed report that gender stereotypes and patrifocal restrictions on behaviour often generate conflict between their personal and professional roles. Not only do these women affirm experiencing social pressure to maintain gender roles within the workplace, they also contend that mobility restrictions limit their informal interaction with colleagues and social networks.

The present study builds on this understanding of patrifocality as a source of professional isolation. Specifically, the cultural emphasis on female purity and family responsibility, as well as the related restrictions prohibiting extensive contact with unrelated men and individual travel, serve to isolate female scientists within their work environments. The impact of this sort of social disconnect is profound in the context of scientific careers, as networking bolsters access to resources that enhance productivity and facilitate collaboration and resource allocation (Lin 2001). In this regard, ICTs have emerged as promising technological solutions to patrifocal limitations on mobility. Theoretically, these innovations should permit interaction and collaboration in a manner that does not violate, and therefore circumvents, patrifocal prescriptions of appropriate female behaviour.

Campion and Shrum's (2004) study of the conditions faced by female scientists in India and Africa provides the basis for this examination. Using data gathered in 1994, when the Internet first began to appear in the developing world, this study sought to locate specific sources of the lower levels of scientific attainment exhibited by these women. The resulting analysis of gender differences in terms of educational and personal backgrounds, research productivity, professional activities, organizational resources, and professional networks, as well as the consequences of these differences, revealed that gender inequality results from a lack of social rather than material resources. Relative to their male colleagues, female scientists in Kerala were found to have lower participation in government committees and editorial boards, publish less in international journals, and possess fewer contacts outside of the local research system, including those with professionals in developed countries. Overall, these findings show the particular influence of patrifocality on the research careers of Indian women in two respects. First, women exhibit higher degrees of educational and research localism. Second, their professional networks are not as wide-ranging as those of their male colleagues. The authors suggest these inequalities can be redressed by making more international opportunities available to female researchers in the developing world. As the same social prescriptions that cause professional isolation also prohibit women from traveling, the feasibility of this solution in the Indian context is somewhat limited.

In the decade since these data were gathered, however, the Internet has emerged as a promising means with which to combat professional isolation among Indian women scientists. Although the benefits of ICT to the social status of developing world women are widely theorized, the assertion that ICT adoption and diffusion will undermine traditional sources of gender inequality remains an untested hypothesis. Pertinent to this analysis is the hesitancy of Indian women to subscribe to western style feminism. Therefore, the issue immediately at hand is the impact of ICT on network size and orientation rather than ability of these technologies to promote feminist consciousness. Before commencing this analysis, I first provide a description of the setting of this study and my methodological approach.

The Kerala Context

From a development perspective, the State of Kerala in southwestern India is, in many respects, a conundrum. The paradoxical pattern of growth, often referred to as the "Kerala Model of Development," has long drawn attention from development communities due to the significant improvements in quality of life achieved in spite of low degrees of economic growth and industrialization (Oommen 1992, 1999; George 1993; Ramachandran 1996; Shrum and Iyer 2000). Due to the Marxist socialist leanings of many of the development initiatives[9] in Kerala and their success,[10] Franke and Chasin (2000) describe the State as "an experiment in radical reform as a modern development strategy (10)."

Perhaps the most debated aspect of Kerala culture is the status of women. Due to traditions of female education and property ownership among Kerala's Nayar caste, the state has gained a reputation for high degrees of gender equality. The celebrated indicators of the status of women in Kerala are the high rate of female literacy and a female to male ratio (1.04: 1), nearly identical to those found in developed nations (Parayil 1996). Critics of these statistical indicators, however, maintain that recent increases in crimes against women and rising rates of suicide and mental illness among women in Kerala reveal another aspect of the status of women (Eapen and Kodoth 2002). Additionally, recent investigation of the matrilineal kinship structures of the Nayar caste reveals that Nayar women held virtually no authority within their households (Renjini 2000).[11]

Methodology

The results below are based on 90 face-to-face interviews conducted between 2003 and 2004 with professional scientists in both university and government sectors in Thiruvananthapuram, Kerala. Out of all respondents, approximately one quarter were female (see Table 1). This gender ratio is slightly lower than the all-Kerala proportion of 37 per cent in 2000 (Sooryamoorthy and Shrum, 2004). The four Indian institutes surveyed included: the Central Tuber Crops Research Institute (CTCRI), and Centre for Earth Science Studies (CESS), both of which are government research stations, and the science departments at the University of Kerala (UK) and Kerala Agricultural University (KAU), two academic centres in Thiruvananthapuram. While nearly all employees at both government institutes were included in this sample, the relatively large size of the universities compelled the selection of a subset of departments in agricultural and basic scientific fields. Women employed in the universities were also purposely oversampled. Although this sampling strategy is not without bias, this analysis is not intended to produce generalizable results. Rather, the goal of this study is to show the manner in which ICTs are beginning to reduce the influence of patrifocality on the careers of female scientists in Kerala.

The interviews were preceded by the review of each respondent's curriculum vitae and recent publications. Transcribed interviews were entered into QSR NVivo, a qualitative data analysis package and coded according to the scheme of "nodes" summarized in Appendix A. Analysis was conducted by way of Boolean searches, which locate passages possessing a particular combination of coded features in the form of intersections, unions, differences, negations, matrix intersections, and matrix differences (Richards 1999). In this analysis, we utilized seven "parent" nodes: **Location of**

Higher Education, Organizational Involvement, Visits Abroad, Professional Contacts, Internet Use and Access, Research Related Email Transactions, and Gender Aspects. In turn, each parent node was broken down into "child" nodes that served to further organize responses in terms of the diffusion of Internet technologies and their impact on female scientists' access to social resources. Searches of the coded interviews were conducted using these nodes and analyzed by sex of respondent. Findings were compared to Campion and Shrum's earlier results to determine the degree to which connectivity in Kerala has increased and how this is affecting the careers of female scientists.

In the following section, I examine the increase in connectivity and Internet access by male and female scientists, including the role of the computer in the domestic setting. Next, I address the two sources of network localism, location of higher education and travel, previously identified by Campion and Shrum as sources of disadvantage. Finally, I speak to the question of circumvention and social change and conclude by discussing the significance of these findings in terms of the larger body of development discourse.

Degree of Connectivity

Although measuring the level of Internet access in Kerala is not the goal of the present study, adequate connectivity is a necessary precondition. In order for female scientists to use ICT to expand their social and professional networks, they must first possess ready access to email and web browsing. As shown by this analysis, Kerala is now a fairly well connected state. Internet connections can be found at all four institutes and a majority of respondents report home connections. Many described improvement in the quality of Internet connections available in Kerala. The majority of scientists in both sectors agreed that the Internet is a useful tool for professional communication and research and most reported regular use and at least basic knowledge of email and literature searching.

A disparity was indicated, however, in terms of degree of connectivity, where governmental research centres were found to possess an advantage over academic institutions. Only employees at CESS and CTCRI felt that they had adequate access to the Internet at work, with two to three computers in each lab. These respondents also expressed satisfaction with the LAN connections recently acquired by the institutes, although those working at CTCRI complained that access was not available outside of office hours except with special permission. As one CTCRI scientist articulated: "Although the connectivity is for 24 hours, but we are able to access it only till office hours. At 3.30 in the afternoon [when office time officially ends] they switch off the main server. Because there are occasions when you need to use it even after office hours – a couple of hours more."

In contrast, university employees reported that they are generally contented with the present arrangement of Internet facilities in their departments, but unhappy with the number of computer systems available and the slow speed of the dial-up connections. Although two or three connections existed in each department, students often monopolize these computers. According to one UK marine biologist, "here in the department only one computer is connected to the Internet. So all the students will be crowding around that all the time. So teachers don't go and use it during office hours".

Despite the availability of Internet access in university libraries and computer labs, nearly all academic scientists reported obtaining home access, and using these connections to carry out the majority of their browsing activities. This use of the Internet in the domestic setting also seems to have particularly benefited female respondents in terms of knowledge of the technology. Simply put, women were more likely than men to attribute their competence using the Internet to their children. This instruction occurred both directly, as when children set up email accounts for their mothers, and indirectly, as a result of supervising Internet use and assisting with homework. When asked where she obtained her Internet knowledge, one female department head responded: "My daughter…and we have [the connection] in our department since about a year…It was difficult as I do not know typing. When I do something, when I try to open or close a window, if something happens I would get stuck. My daughter would help me and then I have a relative close to my house who helped me out".

Several women also reported that they actually depend on their children to use the Internet for them, providing instruction for literature searches and dictating email but never actually touching the keyboard. Several male respondents did mention the influence of their sons and daughters on the decision to purchase a home computer and the instruction they provided once the system had been installed. The respondent who claimed to possess the highest degree of Internet savvy, however, was a man who gained this knowledge while conducting research in developed countries.

Location of Higher Education

Gender differences in level and location of higher education are central indicators of inequality in scientific careers throughout the developing world. As long as women lag behind men in quantity of training and do not have access to the same institutions of training as men, their access to the social capital crucial to the pursuit of scientific careers will be limited. Using the 1994 data, Campion and Shrum found that male scientists in the developing world were more likely to receive training in developed countries, specifically the United States and Europe. Indian researchers, however, were less likely than those from Africa to receive foreign degrees (2004).

Although I did not find the women in my sample disadvantaged in terms of level of education, the location of higher education served as a good indicator of the degree of gender stratification that currently exists in Kerala. The analysis of the location of higher education, particularly doctoral studies, suggests that women were constrained from acquiring their higher degrees from universities outside of Kerala. The vast majority completed their higher degrees, and in many cases all of their degrees, within the State. One female department head accounted for her locally oriented education in the following manner: "We cannot move out freely after the household chores. My husband is not the kind who is supportive...My husband is very orthodox. He would get irritated with studying after marriage. He wanted me to have job but did not like me studying after marriage".

Only one female respondent received a foreign doctorate from an institution. Surprisingly, she selected and attended on her own while her husband and child remained in India. This woman's experience, however, was exceptional due to her independent travel as well as the planning involved in this endeavour. As a group, women did not generally aspire to foreign degrees and many women reported beginning their doctoral studies as a result of having no definite career plans or starting them only after procuring permission from their husbands. Marital values and family obligations further limit women's movements away from their families, functioning as barriers to social and physical mobility.

The situation of male scientists is quite different. For men, whose physical mobility is uninhibited, choice of location of higher education was not an issue. No male respondent spoke of turning down opportunities to study abroad due to family concerns. In contrast, one woman recalled interrupting her studies when her husband accepted a fellowship in Canada

> *My husband was a scientist and he wanted to study. So I told him the same thing. And then he got a fellowship to go to Canada. We went to Canada and he was doing his studies... I also joined in the Life Science department as a research assistant as directly I cannot do any programme and they were doing tissue culture in apple and I learned the technique, working as a project assistant.*

Several women did report having completed their doctoral studies in universities outside of India, but on closer scrutiny, it was discovered that these decisions were inextricably linked to decisions of authoritative males. These women tended to accompany husbands or brothers who first undertook foreign assignments and should not be viewed as having made independent decisions to study abroad. As male and female respondents who studied outside of India reported that professional contacts in these countries continue to serve as collaborators, resource providers, and friends, even foreign studies of this sort are valuable experiences.

Travel Abroad

Research localism, the second source of network localism identified by Campion and Shrum, is measured by days of work related travel (2004). In this previous study, women were found to travel abroad at a lower rate than their male colleagues and spend approximately two weeks fewer per year traveling for conferences, training, or research. The results of the current study indicate that women remain disadvantaged in this respect. Although most of the senior women scientists and professors claimed to have traveled abroad for professional activities, fewer women engage in international travel. Female scientists employed at research institutes tended to undertake more foreign visits than university scientists, but this appears to result from the greater tendency of government research institutions to send their employees abroad for training. Those women who do embark on foreign trips take advantage of more short-term travel opportunities such as conferences, workshops, seminars, and symposia. These international opportunities appeared more plentiful than those that involve teaching or conducting research and to last for shorter periods of time. Conversely, male scientists were more likely to engage in long-term opportunities such as postdoctoral fellowships, research, and employment. These results imply that family obligations still govern the duration of visits abroad for female scientists.

While the current analysis indicates no change over the past decade in number of women traveling abroad, it does reveal a heightened awareness among Indian women of the value of foreign travel. Female respondents who had traveled to other countries spoke in depth about the professional ties that resulted from these trips and the importance of presenting their work to international audiences. Several women also recounted discovering international opportunities and going to great lengths to fund their travels. Due to the superior resources often available through ties with scientists in the developed world, and the decreased likelihood that these individuals subscribe to beliefs that restrict contact between men and women, the importance of traveling abroad for Indian scientists cannot be overstressed. If women are still unable to regularly travel abroad or to do so for significant periods of time, they are denied the opportunity to develop ties that can effectively circumvent patrifocal restrictions on their behaviour. In the absence of these face-to-face interactions, the question arises as to whether Internet-based interaction can cultivate similar relationships.

Professional Contacts

The importance of international travel and research to scientific careers becomes more apparent in light of the network ties that result from such undertakings. Although social restrictions on behaviour and mobility have long disadvantaged Indian women scientists in this regard, electronic communication made possible by new ICT has emerged as a possible equalizing mechanism. Campion and Shrum's (2004) finding that female scientists possess larger networks within Kerala while male scientists possess larger Indian networks outside Kerala, provide the foundation for this inquiry. To the extent that Internet-based communication can substitute for face-to-face contact, these technologies should allow Indian women to circumvent social restrictions and cultivate larger, less locally oriented networks. Although the qualitative interviews utilized in the current study do not allow a confirmation of this hypothesis, they provide exciting evidence of social change occurring over the past decade.

As this analysis reveals, Indian women discussed their various professional contacts throughout the world, and, contrary to expectations based on Campion and Shrum, mentioned far fewer local contacts than Indian men. Instead, they were more likely to describe their associations with various international scientists. One woman working in the academic sector reported that her use of international contacts had begun while she was still completing her PhD research: "While doing PhD we used to contact and found out different people from different countries…Not much within India…I have certain friends in Japan…in Ireland…Czech Republic… Canada…Poland." Women employed in the government sector seemed to list more local male contacts than their university counterparts, but many of these men occupied supervisory positions and were expected to provide feedback. Similarly, the Indian male contacts of female respondents were more often identified as lab partners, former professors, or dissertation advisors than scientists who sought them out on the basis of previous publications or for the purpose of collaboration. Contacts of these types were generally not used for collaboration, but rather for examination purposes or settling questions. In contrast, Indian men reported extensively on contacts and collaborations within their departments, institutes, and regional areas.

Many female respondents claimed that the Internet was responsible for the cultivation of their international ties. Several mentioned submitting papers to international journals or conferences about which they received information online. Also common were accounts of publishing papers in online journals and subsequently receiving email correspondence from interested readers. According to one UK marine biologist:

> *Yes, I used to get some emails and all, for my papers; some from, abroad some from Iran. In that case we use to send them papers. We too used to get papers for references. Like that people from Pakistan, Iran, and all used to send mail requesting papers and of course while I was doing study on pesticides I used to get lot of mail from environmental protection agency, USA.*

In a few cases, such exchanges led to visits abroad and international collaboration. Additionally, all participants in the study claimed that email was their preferred mode of communication with all of their contacts, although some expressed the preference of sending letters by way of post to those professional contacts that were also friends. Overall, the utility of email as a tool to enhance productivity was widely accepted despite the fact that it was also widely held to be more impersonal than communication by telephone or post.

Discussion

As shown by these qualitative interviews, the Internet is beginning to play a rather noteworthy role in the careers of Indian women researchers. The diffusion of the Internet in Kerala is resulting in the circumvention of the patrifocal social structure that restricts the physical and social mobility of women (Palackal et al. 2006, 2007). Although this hypothesis should not be interpreted too liberally, several key changes in the careers of Indian women scientists do lend support to the premise that technological advancement can result in social change.

In terms of social networks, I have found that Indian women scientists *should* possess more locally oriented ties than their male colleagues. They are more likely to obtain higher degrees from institutions within the state in which they currently reside, less likely to embark on research related trips abroad, and, when they do travel abroad, less likely to spend long periods away from home. In these respects, Campion and Shrum (2004) and Gupta and Sharma (2002) are likely to be correct in their assessments of the social networks of Indian women scientists. However, the Internet seems to have impacted the resource acquisition capabilities of Indian women in a manner that has allowed them to begin to overcome their professional isolation, at least in terms of two aspects of awareness important for professional career development. First, female respondents discussed more international professional contacts than do males. Second, over the past decade, Indian women seem to have gained a heightened awareness of the importance of international travel. Essentially, Internet technologies have allowed Indian women access to social and organizational resources outside of the Indian patrifocal setting and have increased their consciousness of the international scientific community.

These findings should not be interpreted in an overly optimistic fashion, as the patrifocal social structure seems firmly in tact. Indian women scientists do not seem to have become more globally oriented in terms of education or travel. Rather, the area in which change is revealed is that of the local orientation of social networks. Indian women were found to have discussed their international contacts more than their male colleagues, but whether or not they actually possess more global network ties is impossible to infer on the basis of this analysis.

Despite the necessary caution exercised in this interpretation, one final dimension of these results give cause for optimism. The fact that Indian women spoke of their international contacts on a more frequent basis than did men implies that they place a higher value on these network ties and recognize the benefits of networking outside of the patrifocal social context. Ultimately, the Internet seems to be providing Indian women access to social and material resources previously beyond their reach and this increased access may also be functioning to undermine culturally entrenched practices of gender stratification. Although quantitative investigation of network size and composition in terms of age, sex, and location of respondents is warranted before more definite conclusions can be drawn, this investigation has established a satisfactory starting point for further study. We may never be able to determine the extent to which, independent of other western influences, ICTs function to undermine the Indian patrifocal social structure. This analysis, however, has provided evidence that Indian women scientists have taken advantage of the social and professional opportunities made available to them by the diffusion of these technologies.

In response to recent criticism of ICT-driven development policy as incompatible with local context and, therefore, incapable of promoting sustainable development, I offer the results of this analysis. By confining the scope of this study to professional scientists in Kerala, India, I was able to assess the impact of ICT in a manner consistent with the sociocultural climate as well as the particular needs and abilities of the respondents. I have located evidence of the benefits of Internet adoption without resorting to overarching explanations or homogenized accounts of gender stratification in the developing world. Although my sample is a relatively advantaged slice of the population, the benefits they have garnered from technological adoption show its utility in some developing world contexts. Through the use of Internet technology, female scientists in Kerala appear to have begun to effectively circumvent patrifocal limitations on their physical and professional mobility. Rather than passive recipients of western development, these women can be seen as active agents in the erosion of patrifocality.

Notes

1. The essay is closely related to the following stream of work by the World Science Project: Meredith Anderson and W. Shrum. 2007. "Circumvention and Social Change: ICTs and the Discourse of Empowerment". *Women's Studies in Communication*. (Forthcoming); Miller et al 2006, Palackal et al. 2006, Palackal et al. 2007. This is the first previously unpublished article in the series. This material is based upon work supported by the National Science Foundation under Grant No. 0113545.
2. Campion and Shrum attributed the lower degrees of attainment observed in the careers of female scientists in Kerala to systematic deficits in social resources evidenced in higher degrees of educational, research, and network localism than their male colleagues.
3. The thrust of this policy, which aims to eliminate gender inequality in primary and secondary education by 2005, and in all levels of education by 2015, is that equal access to all levels of education for women spurs social, economic, and political empowerment. Additionally, an increase in female education would facilitate the realization of other Millennium Development Goals, including the reduction of child mortality, improving maternal health, and the prevention of HIV, malaria, and other diseases.
4. Wajcman cites the example of the microwave, a product developed for single men but widely adopted by housewives. The unintended use of this technology ultimately resulted in changes in its appearance and social significance.
5. The social action surrounding the adoption and diffusion of the Pap smear illustrates how individual- and organizational-level activism can shift the social construction of a technology. Casper and Clark (1998) describe how activism spearheaded by the American Cancer Society changed the social perception of the technology from inaccurate to precise. Likewise, Wajcman (2004) recalls later feminist action that served to shift the social construction of cervical cancer screening from a de skilled and feminized field to an important use of technology.

6. This list is in no way complete, as many other works on the topic have been published in recent years. See Fox (1999) for a review.
7. Escobar and Harcourt (2005) maintain that patriarchal power relations also exist in the developing world as a result of globalization, industrialization. Although the form of western patriarchy varies from place to place, it is an omnipresent force, often resulting in increasingly complex and unequal power relations between women and men, among women and among men.
8. A good example of this type of power dynamic is the authority exercised over younger women by high-ranking women such as mothers-in-law.
9. Examples of these programmes include the 1969 land reform, in which the abolition of tenancy resulted in the redistribution of land rights, and public health programmes featuring the redistribution of food and government subsidized "ration shops."
10. According to Franke and Chasin (1994), Kerala's social development initiatives have resulted in a myriad of impressive social statistics including the lowest infant mortality rate, the lowest rate of population growth, and the highest levels of adult literacy in all of India.
11. Although Nayar women legally controlled all property belonging to the household, its management was undertaken by the *karanavan*, the senior males members of the kin group. Essentially, the *karanavan* exercised total authority over all women and children in the household. Nayar women generally had casual relationships with their husbands, who maintained separate residence from their wives and held little authority over them. Due to the authority wielded by the *karanavan*, however, Nayar women were actually subject to the same system of patrifocality as women in other land owning castes.

Appendix A

Thematic Organization Used in Interview Coding

LOCATION OF HIGHER EDUCATION

Post Graduation
- institute

Doctoral Studies
- institute
- period of commaencement

— area of study
— findings of the study

Post Doctoral Study

— institute
— period of commencement
— area of study
— findings of the study

Others

ORGANIZATIONAL INVOLVEMENT

— Professional
— Non Professional
— No Involvement

VISITS ABROAD

Training/Workshops
Fellowship
Guest Faculty
Obtaining Degrees/Taking Courses
Conferences
Deputation
Resource Person
Effect/Influence
No Professional Visits Abroad

PROFESSIONAL CONTACTS

Purpose
— professional
— non-professional
Origin of the Contact
Level of Contact
Mode of Communication

INTERNET USE AND ACCESS

Impression About Internet
Latest Internet Activity
Problems Encountered
Unique Internet Experiences
Introduction to Computers
Impact of Internet

— teaching
— social life
— research
Use of Internet
— research
— teaching
— others
Browsing
— duration
— sites visited
— ways of browsing
Access to Internet
— institute
  — speed
  — availability
  — future developments in Internet access
— home
  — speed
  — availability
— café
  — speed
Negative Impact of Internet

EMAIL

Email Activity/Duration Rate
Nature of Email Response
Correspondence Built on Email
Rate of Mails sent and Received
Email Transactions
— partners
  — personal
  — professional
    — research related
    — others

GENDER ASPECTS

Research
Occupation
Study

References

Abbate, Janet. 2000. *Inventing the Internet*. Cambridge, MA: MIT Press.

Abraham, Margaret. 2000. *Speaking the Unspeakable*. New Brunswick: Rutgers University Press.

Acosta-Belen, Edna and Christine E. Bose. 1990. "From Structural Subordination to Empowerment: Women and Development in Third World Contexts," *Gender and Society* 4(3): 299-320.

Boserup, Esther. 1970. *Woman's Role in Economic Development*. New York: St. Martin's Press.

Campion, Patricia and Wesley Shrum. 2004. "Gender and Science in Developing Areas," *Science, Technology, and Human Values* 29 (4), 459-85.

Casper, Monica J. and Adele E. Clark. 1998. "Making the Pap Smear into the 'Right Tool' for the Job: Cervical Cancer Screening in the USA, circa 1940-95," *Social Studies of Science* 28(2), 255-290.

Chakravathy, Radha. 1986. "The Productivity of Women Scientists," *Productivity* 27 (3), 259-269.

Chanana, Karuna. 2001. "Hinduism and Female Sexuality: Social Control and Education of Girls in India," *Sociological Bulletin*, 50 (1), 37-63.

Cole, Jonathan R. and Harriet Zuckerman. 1984. "The Productivity Puzzle: Persistence and Change in Patterns of Publication of Men and Women Scientists." in *Women in Science*. Marjorie W. Steinkamp and Martin L. Maehr (eds.). Greenwich and London: JAI Press, 217-258.

Davidson, Theresa, R. Sooryamoorthy, and Wesley Shrum. 2002. "Kerala Connections: Will the Internet Affect Science in Developing Areas?" Pp. 469-519 in: *The Internet in Everyday Life*. B. Wellman and C. Haythornthwaite (eds.). Malden, MA: Blackwell.

Derne, Steve. 1994. "Arranging Marriages: How Father's Concerns Limit Women's Educational Achievements" in *Women, Education, and Family Structure in India*. Carol C. Mukhopadhyay and Susan Seymour (eds.). San Francisco: Westview Press, 83-101.

DoT Force. 2001. "Digital Opportunities for All: Meeting the Challenge". New York: United Nations Digital Opportunities Task Force, Retrieved February 10, 2006 (lacnet.unicttaskforce.org/Docs/Dot%20Force/Digital%20Opportunities%20for%20All.pdf).

Drucker, Peter. 2001. "The Next Society". *Economist*. 361 (8264), 3-5.

Eapen, Mridul and Praveena Kodoth. 2001. "Demystifying the 'High Status' of Women in Kerala – An Attempt to Understand the

Contradictions in Social Development". Working Paper 341. Thiruvananthapuram: Centre for Development Studies.

Escobar, Arturo. 1995. *Encountering Development: The Making and Unmaking of the Third World*. Princeton, NJ: Princeton University Press.

Escobar, Arturo and Wendy Harcourt. 2005. "Practices of Difference: Introducing Women and the Politics of Place" in *Women and the Politics of Place*. Wendy Harcourt and Arturo Escobar (eds.). Bloomfield, CT: Kumarian Press, 1-19.

Fox, Mary Frank. 1995. "Women in Scientific Careers" in *Handbook of Science, Technology, and Society*. Sheila Jasanoff, Gerald Markle, James Peterson and Trevor Pinch (eds.).Newbury Park: Sage, 205-223.

—. 1999. "Gender, Hierarchy, and Science". In *The Handbook of the Sociology of Gender*. Janet Salzman Chafetz (ed.). New York: Kluwer Academic/Plenum Publishers, 441-458.

Fox, Mary Frank and Scott J. Long. 1995. "Scientific Careers: Universalism and Particularism," *Annual Review of Sociology*. 50, 45-71.

Franke, Richard W. and Barbara H. Chasin. 2000. "Is the Kerala Model Sustainable? Lessons from the Past, Prospects for the Future" in *Kerala: The Development Experience*. Govindan Parayil (ed.). London: Zed Books, 16-39.

Gajjala, Radhika. 2003. "South Asian Digital Diasporas and Cyber Feminist Webs: Negotiating Globalization, Nation, Gender, and Information Technology Design," *Contemporary South Asia* 12 (1): 41-56.

Gajjala, Radhika and Annapurna Manidipudi. 1999. "Cyberfeminism, Technology, and International Development," *Gender and Development* 7(2): 8-16.

George, K. K. 1999. *Limits to Kerala Model of Development: An Analysis of Fiscal Crisis and its Implications*. Thiruvananthapuram: Centre for Development Studies.

Gupta, Namrate and Arun K. Sharma. 2002. "Women Academic Scientists in India," *Social Studies of Science*. 32 (5-6), 901-15.

Harcourt, Wendy. 2005. "The Body Politic in Global Development Discourse: A Woman and the Politics of Place Perspective" in *Women and the Politics of Place*. Wendy Harcourt and Arturo Escobar. (eds.). Bloomfield, CT: Kumarian Press, 32-47.

Harding, Sandra. 1995. "Just Add Women and Stir?" in *Missing Links*. eds Gender Working Group: UN Commission on Science and Technology for Development. New York: UNIFEM, 295-307.

Kelkar, Govind and Dev Nathan. 2002. "Gender Relations and Technological Change in Asia". *Current Sociology*. 50 (3), 427-441.

Keller, Evelyn Fox. 1995. "The Origin, History, and Politics of the Subject Called 'Gender and Science': A First Person Account". In Sheila Jasanoff, Gerald Markle, James Peterson, and Trevor Pinch (eds.). *Handbook of Science, Technology, and Society*. Newbury Park: Sage, 80-94.

Kyvik, Sven and Mari Teigen. 1996. "Child Care, Research Collaboration, and Gender Differences in Scientific Productivity." *Science, Technology, and Human Values* 21 (1), 54-71.

Lin, Nan. 2001. *Social Capital*. Cambridge: University Press.

McElrath, Karen. 1992. "Gender, Career Disruption, and Academic Rewards". *The Journal of Higher Education* 63 (3), 269-81.

Meredith Anderson and W. Shrum. 2007. "Circumvention and Social Change: ICTs and the Discourse of Empowerment". *Women's Studies in Communication*. (Forthcoming).

Mies, Maria. 1986. *Patriarchy and Accumulation on a World Scale: Women in the International Division of Labour*. London: Zed Books.

Miller, B.P., R. Sooryamoorthy, M. Anderson, A. Palackal, W. Shrum. 2006. "Gender and Science in Developing Areas: Has the Internet Reduced Inequality". *Social Science Quarterly* 87 (3), 679-689.

Mohanty, Chandra Talpade. 2003. *Feminism without Borders*. Durham, NC: Duke University Press.

Morgall, Janine Marie. 1993. *Technology Assessment: A Feminist Perspective*. Philadelphia: Temple University Press.

Mukhopadhyay, Carol C. and Susan Seymour. 1994. *Women, Education and Family Structure in India.* Boulder: Westview Press.

Oommen, M. A. 1992. *The Kerala Economy*. New Delhi: Oxford and IBH Publishing Company.

—. 1999. *Kerala's Development Experience* (Vols. 1 and 11). New Delhi: Concept Publishing.

Pal, Joyojeet. 2003. "The Developmental Promise of Information and Communications Technology in India". *Contemporary South Asia*. 12(1), 103-119.

Palackal, A., M. Anderson, B. P. Miller, and W. Shrum. 2006. "Gender Stratification and E-Science: Can the Internet Circumvent Patrifocality"? in C. Hine (Ed.), *New Infrastructures of Knowledge Production: Understanding E-Science*. Hershey & London: Information Science Publishing. 246-271.

Palackal A, M. Anderson, B. P. Miller and W. Shrum. 2007. "Internet Equalizer? Gender Stratification and Normative Circumvention in Science", in *Indian Journal of Gender Studies*, 14 (2).

Parayil, Govindan. 1996. "The 'Kerala Model' of Development: Development and Sustainability in the Third World," *Third World Quarterly* 17 (5), 944-957.

Park, Kyung Ae. 1993. "Women and Development: The Case of South Korea," *Contemporary Politics* 25 (2), 127-145.

Patel, Reena and Mary Jane C. Parmentier. 2005. "The Persistence of Traditional Gender Roles in the Information Technology Sector: A Study of Female Engineers". *Information Technologies and International Development* 2 (3), 29-46.

Ramachandran, V. K. 1996. "On Kerala's Development Achievements." in *Indian Development: Selected Regional Perspectives*. J. Dreze and A. Sen (eds.). Delhi: Oxford University Press, 205-356.

Ramu, G.N. 1987. "Indian Husbands: Their Role Perceptions and Performance in Single- and Dual-Earner Families", *Journal of Marriage and the Family* 49 (4), 903-915.

Ranson, Gillian. 2003. "Beyond 'Gender differences': A Canadian Study of Women's and Men's Careers in Engineering," *Gender, Work, and Organizations* 10 (1), 22-41.

Renjini, D. 2000. *Nayar Women Today: Disintegration of Matrilineal System and the Status of Nayar Women in Kerala*. New Delhi: Classical Publishing Company.

Richards, Lynn. 1999. *Using NVivo in Qualitative Research*. London: Sage.

Riessman, Catherine Kohler. 2000. "Stigma and Everyday Resistance Practices: Childless Women in South India," *Gender and Society* 14 (1), 111-135.

Rout, Usha Rani, Sue Lewis, and Carolyn Kagan. 1999. "Work and Family Roles: Indian Women in India and the West," *Indian Journal of Gender Studies* 6 (1), 91-104.

Shrum, Wesley. 2005. "Reagency of the Internet, or How I Became A Guest for Science," *Social Studies of Science*. 35 (5), 723-754.

Shrum, Wesley and S. R. Iyer. 2000. "Knowledge, Democratization, and Sustainability: The Kerala 'Model' of Scientific Capacity Building" in *Kerala: The Development Experience*. Govindan Parayil (ed.). London: Zed Books, 151-177.

Shukla, Archana. 1987. "Decision Making in Single- and Dual-Career Families in India," *Journal of Marriage and the Family* 49 (3), 621-629.

Sooryamoorthy, R. and Wesley Shrum. 2004. "Is Kerala Becoming a Knowledge Society? Evidence from the Scientific Community," *Sociological Bulletin* 53 (2), 207-221.

Subrahmanyan, Lalita. 1998. *Women Scientists in the Third World: The Indian Experience*. New Delhi: Sage.

United Nations. 2005. "The Millennium Development Goals Report 2005". New York: United Nations.

Wajcman, Judy. 1991. *Feminism Confronts Technology*. University Park, PA: The Pennsylvania State University Press.

—. 1995. "Feminist Theories of Technology" in *Handbook of Science, Technology, and Society*. Sheila Jasanoff, Gerald Markle, James Peterson, and Trevor Pinch (eds.).Newbury Park: Sage, 189-204.

—. 2004. *Technofeminism*. Cambridge: Polity Press.

Xie, Yu and Kimberlee A. Shauman. 1998. "Sex Differences in Research Productivity: New Evidence About an Old Puzzle," *American Sociological Review* 63(6), 847-70.

Youngs, Gillian. 2002. "Feminizing Cyberspace" in *Rethinking Empowerment: Gender and Development in a Global/Local World*. Jane L. Parpart, Shirin M. Rai, and Kathleen Staudt. eds. London: Routledge, 79-94.

10

Knowledge Production and Developmental Structures

E-science in a Developing World

ANTONY PALACKAL AND WESLEY SHRUM

The purpose of this paper is to address whether the Internet changes the practice of science in less developed areas by examining its impact on the research careers of scientists. It opens up a line of analysis on the nature of scientific knowledge and developmental structures in the State of Kerala after the initial diffusion of ICTs. The larger question is whether the diffusion of ICTs enables scientists in less developed areas like Kerala any advantages in producing knowledge. Many scholars have argued that Kerala is a unique setting in the developing world, especially when examined on the basis of the introduction of the Internet (Franke and Chasin 1991, 2000; Parayil 1996). The results of this paper are based on a field study conducted among scientists in research and academic institutes in Kerala in the year 2005. The findings of this empirical survey are examined in the light of a prior study conducted in the same field. Hence this paper employs a longitudinal framework juxtaposing its results with the analysis and conclusions of an earlier article by Ynalvez et al. (2005) on fieldwork carried out in 2000. We intend to make a diachronic analysis of the degree to which scientists in the developing world have adopted the ICTs that are now a prerequisite for participation in global scientific endeavours.

The assumption that ICTs have the potential to make societies increasingly knowledge based constitutes the substratum of the study. Very often an argument is made in the academic parlance that present day scientific developments are based on innovations in ICTs. It has even been argued that the diffusion of ICTs in developing areas will lead to the globalisation of science, furthering the universal benefits of ICTs (Robinson 2003). This will in turn lead to integration, participation, and visibility of developing world scientists in the global scientific enterprise. Therefore this study seeks to answer the question: "Does new communication technology change the nature and structure of science practices in Kerala"? Scientists in the developing countries have for decades been perceived as isolated, a state of affairs with both informational and interpersonal dimensions (Dedijer 1963; Salam 1966). In developing areas acquisition costs for necessary and appropriate materials and literature for research are often unaffordable and scientists are limited in the form of inadequate libraries and documents. The interpersonal level of isolation takes the forms of barriers in contacting active research communities in the West. In this context, the basic question raised in this paper is whether the Internet will be a solution to the problem of isolation?

Lamb and Davidson (2005) provide a typology of the use categories of ICTs by scientists: embedded, coordination and dissemination. Embedded ICTs are communication tools built into scientific tools and experiments such as sensor networks, grid computing, and remotely-operated telescopes and observation devices, visualization and virtual reality tools, and telemedicine tools. They are primarily part of collaboratories and make "big science" possible (Kauzes 2000). Coordination ICTs such as email, telephone, web pages, instant messaging, chat, wikis and so forth constitute the general communication information infrastructure that allows scientists to plan, share data and results, write papers and maintain contacts. Dissemination ICTs such as electronic journals, popular media, weblogs and project websites transmit the findings to the audience in one way broadcasts (Lamb and Davidson 2005). Liebscher et al (1997) studied the use of ICTs in smaller institutions and found that institutional support was an important predicator of use. In 2003 a survey of scientists in research and development organisations in seven European countries found that 97.7 per cent of scientists reported using email (Barjak 2004a). However, Russell (2001) states that scientists in third world countries have been slower to adopt ICTs because of the lack of institutional infrastructure, power and telecommunications.

The Internet is of special significance as it has been a powerful feature in the information area since its inception in the last quarter of the 20th century. Today 'information' is the most vital resource for any kind of activity. The Internet has quickly become a major player in the quest for information. People readily employ the Internet as a tool for communication and information retrieval because it can provide fast access to vast amounts of information from a multitude of sources. The Internet and email expand and strengthen the social ties that people maintain in the offline world. With the help of the Internet, people are able to maintain active contact with sizable social networks with distant others. Further, the Internet provides access to valuable resources scattered in various forms in different parts of the world. An Internet user has access to a wide variety of services: electronic mail, file transfer, vast information resources, interest group membership, interactive collaboration, multimedia displays, real-time broadcasting, shopping, opportunities, breaking news, and much more. Though restricted in its initial stages of development, in the 1990s the content and use of the Internet expanded to include many areas such as business, industry, education, research, government and entertainment, etc.

Science and the Internet

The Internet has emerged as an essential tool in the scientific practice of the developed world. It has come to be known as the information superhighway (Mulla and Chandrasekara 2006). Since the Internet combines a diversity of unique features, including interactional components as well as content formats, its impact on social change is likely to be distinct from other technological innovations such as the television and the telephone (DiMaggio et al. 2001). Modern Internet technology has its origins in the development of means for more efficient scientific communication and holds greater promise of integrating the global research community than any previous technology (Davidson et al. 2002). As the Internet becomes a pre-requisite for research and international collaboration, it is important to understand its impact on the scientific community in the developing world.

In recent years, the Internet has changed the whole process of information handling. It serves as a backbone and connects to the sources of information irrespective of their locations and has taken all the responsibilities of controlling the problems like collection, organising, storing, retrieval and dissemination of information. Two separate dimensions of the Internet may be distinguished: interpersonal and collective communication practices as well as information exchange and retrieval. It is argued that as a technology, the Internet is basically regarded as a cultural construct endowed with cultural traits of its environment as well as embedded in different cultural spaces (Uimonen 2001). This view emphasises that place, identity and time are important factors in the construction of social realities of people. We have employed the concept of 'reagency' to understand the effects of the Internet on scientific endeavours. Shrum (2005) introduced the concept for the first time in the context of examining the conceptual and substantive roles of the Internet in globalizing science. Reagency emphasizes the unpredictable reactions brought about by the variable nature of time, place, and identity in the information age. Co-presence in the information age is not necessarily a criterion for communication between actors located in geographically dispersed areas. Due to the collaborative nature of the Internet, it has the potential to reagentise science in terms of its capacity to shift the relations between actors in fundamentally different ways than prior development projects (Shrum 2005).

The Internet is regarded as reagentive in the sense that it is an essential technology in generating, producing and disseminating scientific knowledge. A reagent has the capacity to induce a reaction and bring about subsequent changes in an existing phenomenon. It is assumed that Internet can reagentise science in developing areas more effectively than prior developmental initiatives (ibid). The working premise in this perspective is that development initiatives of all kinds create a response, that is, a reaction on the part of the actors involved that is uncontrollable by the agents of development by setting in motion chains of events. This perspective thus accounts for the temporal orientations of actors possessing distinct identities and situated in specific structural contexts. In this study, therefore we ask specifically whether or not the Internet will change the practice of science in Kerala by connecting people working in different regions of the world, fostering communication and collaborations not bound by time and space?

Science Communication

As an enterprise, science is an inherently social and communal activity. Scientists tend to collaborate with their colleagues in the sharing of information and ideas, the publishing of articles, and the evaluation of each other's work. In this sense, computer networks are also social networks. Capital, including social capital in the form of research contacts is becoming more and more global in character, spanning geographic boundaries. Several authors therefore emphasize the importance of communication to science. Garvey (1979) for instance, states: "communication is the essence of science" and so scholarly communication forms the backbone of science practice that connects scientists and enables scientific progress.

Although information and communication technologies (ICTs) have transformed the world in many ways, scholarly scientific communication forms a socio-technical interaction network in which communication is influenced by technology but defined by the social structures of scientists and their organisations (Kling, McKim, and King 2003). A review of studies on the nature of scientific communication and the influences of information communication technologies on existing structures in the developed world reveal that these technologies have not changed the social structure of science, but have enabled new forms of remote collaboration and, perhaps, higher productivity as measured by number of publications. Scientists communicate to brainstorm ideas and be creative, formulate research questions and solve experimental or theoretical problems, disseminate results and get feedback.

Formal scholarly communication in science is published material that has been reviewed by peers, edited by publishers and is retrievable through various information systems (Garvey and Griffith 1972; Sondergaard et al. 2003). Scientists rely on published journal articles for recognition, reward and identity formation. The goal of the formal process is to publish a journal article in a peer reviewed journal. As opposed to popular science communication, scholarly communication is embedded in the context of the scholarly tradition of the discipline and is shaped by disciplinary rituals and perspectives (Fry 2006). Garvey and Griffith (1967, 1972) provided the standard model of the flow of scientific information that stands as the basis for understanding the timeline and milestones for scientific communication. They trace the communication processes from the initiation of the work through the publication of the polished report in a peer-reviewed journal, a process that can extend to five years. From the communication point of view, the results of scientific research are presented in increasingly more formal and vetted outlets over time, gaining in authority as the information passes peer review and editorial changes, gaining larger audiences more distant from the researcher, but losing in immediacy, specificity, and thoroughness (Lievrouw and Carley 1990).

A competing model of scholarly communication is the UNISIST model published in 1971 by the United Nations Educational and Scientific and Cultural Organisation (UNESCO) and the International Council of Scientific Unions (ICSU) (Allen 1977; Sondergaard et al. 2003). While the Garvey and Griffith model emphasises the timeline of communication, the UNISIST model emphasises three channels of communication – formal, informal and tabular (i.e., scientific data rather than text); and three levels of sources – primary, secondary (including library catalogues and abstracting and indexing services) and tertiary (such as reviews and encyclopaedias). These channels and sources connect the information producers and information users.

By contrast, informal communication can take place at any time, anywhere and in any format. Besides communicating for advice, to learn about new methods or theories, or to hear new results, scientists communicate informally to collaborate on research, co-author formal publications and also to gossip and be creative (Kasperson, 1978 cited in Allen 1991). Ideas diffuse more quickly via informal communication than through journal articles alone as they have champions who can provide subjective details on the innovation. Besides actively searching for information to formulate the research question or get feedback on research recently completed, scientists communicate informally in order to meet future research partners, to keep abreast of their research area and to facilitate serendipitous discoveries of information. Hence, one composite definition of informal communication includes interactive exchange of information between scientists to establish or maintain relationships, exchange scientific information, or work collaboratively.

The traditional model of scholarly scientific communication emphasizes the individual production of knowledge with feedback from colleagues and subsequent exchange of knowledge. In other words, interaction is limited to certain stages of research projects, or to proximate researchers (Glaser 2003). Early discussions of ICTs in Science included optimistic prognostications of increased productivity, the end of peripherality (Gresham John Jr. 1994) – in short, the end of the scientific communication structures described by Garvey and Griffith (1972) and UNISIT model. Crane (1972) studied the citation structures and formal publication patterns to describe the growth of science and the "informal but specialized social system that produces basic scientific knowledge" (1). Crane (1972) found that social circles of researchers forum who have similar research interests and are using the same paradigm but are not necessarily co-located. These communication networks, sometimes forming around a scientific paradigm, are called "invisible colleges".

Traditionally informal scholarly scientific communication is carried out through face-to-face meetings, letters and pre-prints. Garvey and Gottfredson (1977) surveyed 2000 scientists and found that the most likely source of research communication was face-to-face contact (cited by Lacy and Bush 1983). Literature on scholarly communications via traditional channels listed peripherality as a concern and limitation on the growth of science. Peripheral scientists are socially or geographically distant and are not tightly integrated into the invisible college, missing opportunities for information sharing and social contact (Hesse et al. 1993). Younger scientists and scientists at smaller universities and scientists in non-Western countries fall into this category. Barjak (2004a) notes that the divide between central and peripheral scientists mentioned in the early literature has transferred to the "Internet Age" within developed Western countries. It is our contention that the concept of the digital divide is more likely to move away from the simple inequalities in hardware-software access and use and moves toward the more complex inequalities of hardware-software-user interaction skills, which are a reflection of the surrounding social capital.

Collaboration and Productivity

Possible returns to adoption and use of ICTs in science are increased productivity, decreased peripherality and broader and more geographically dispersed collaborations. Defining the productivity of scientists is not trivial. Kaminer and Braunstein (1998) and Walsh and Maloney (2002) measured productivity in different ways. A simple measure is to count the journal publications produced, which does not, however, take into consideration the quality of publications, individual contribution of co-authors, other venues of publication, funding received and so forth (Barjak 2005). Hesse et al. (1993) found that productive scientists are more frequent and more powerful users of the network than non-productive scientists. They could not however determine the direction of causality from their study. Kaminer and Braunstein (1998) found Internet use does correlate with higher productivity. In 2003, 20 per cent of all articles indexed in ISI's web of science had authors from multiple countries, a dramatic increase from the 8 per cent in 1998 (National Science Foundation, Division of Science Resource Statistics 2006). Additionally, the average number of authors per article grew from 3.06 in 1998 to 4.22 in 2003 (National Science Foundation, Division of Science Resource Statistics 2006). Research links some of these results to ICT adoption and use. ICTs enabled more extensive collaboration by reducing costs, replacing travel and expensive telephone conversation with low cost ICT communication (Lievrouw and Carley 1990). Wash and Bayman's (1996a) study participants reported increases in the size of research groups and an increase in remote collaborations with the addition of computer networks.

Collaboratories are a new phenomenon centring on the creation of large research projects involving geographically dispersed but ICT-connected researchers (Finholt and Olson 1997). Collaboratories present a social structure fundamentally different from Crane's (1972) invisible colleges because participants are brought together by common equipment and technology, not necessarily by similar research backgrounds and goals. As discussed above, ICTs have impacted the ability of the scientists to collaborate remotely; accordingly, the number and geographic dispersion of collaborations have increased dramatically. Scientists also use ICTs for communications outside of research for information search and maintenance of social contacts. Walsh et al (2000) found that participants perceived that increased email use contributed to increased contact with researchers in other institutions, increased information about conferences, and greater awareness of calls for papers. However, we submit that the shift to Internet-based science in

Kerala, although it enables scientists to circumvent global isolation in general, does not necessarily erase differences such as those associated with gender.

The analysis below elaborates Internet use and access patterns for a sample of scientific and educational institutions in Kerala. First we describe the sample and method. After describing the availability of computers, email access, and web use in some detail, the social structural characteristics associated with these technologies are examined with a view to map change over time in the conditions of knowledge production and scientific practice in Kerala.

Methodology

The participants in our study belonged to two organizational types that are engaged in the conduct of research: (1) the governmental sector and (2) the university sector. The sample of scientists was drawn from these two sectors within each of three study locations. Face to face surveys were administered to 261 scientists belonging to a variety of research fields such as agriculture, biological science, engineering, mathematics, information technology, chemistry, physics, geology in Trivandrum, the capital of Kerala. The questionnaire consisted of approximately 200 questions on social and demographic background factors, professional activities, organizational structure, access to technology, and Internet use.

The interview schedule contained six independent dimensions relating to Internet adoption and several control variables indicating the social, educational, and professional characteristics of respondents. Internet adoption measures include dichotomous measures of ready access to email (1=access; 0=no access) and current email use (1=current user; 0=nonuser), two indicators of Internet experience (years of email and years of web use), and two indices of Internet practice (email use diversity and web use diversity). Email diversity is constructed as the sum of dichotomous (1=yes; 0=no) responses to six questions on the use of email.[1] Web diversity is constructed as the sum of thirteen dichotomous (1=yes; 0=no) items on the use of the World Wide Web.[2]

We present these preliminary results in the following order. First, we highlight certain significant elements concerning the demographic profile of the respondents. In the second part we present data on scientists' access and use of personal computers, email, and the web. We then show differentials in email use, ready access, email and web use diversity, and Internet experience by a variety of social and professional characteristics. Finally, we hint at possible changes in scientific productivity in local and foreign journals between the two study phases.

Profile of the Scientists – Highlights

The gender ratio favours men in both samples. Men generally outnumbered women both in the academic and research sectors, except in the academic sector in 2000, where female scientists outnumbered males. Interestingly, in both periods the gender balance remained more or less the same. The percentage of males in 2000 and 2005 was 62.7 per cent and 62.5 per cent respectively. Almost all respondents were married and, though the difference is negligible, there were more unmarried men (five each in both the periods) than unmarried women (only one in 2000 and two in 2005).

Considerable variation was observed as regards those who secured PhD, the most common advanced degree. The proportion of females with a doctorate was greater than males, and there was a corresponding reduction in the percentage of females with masters as their highest degree. Female scientists seem to have secured more educational capital in terms of the highest degree in 2005 than their male colleagues. However, marked differences were noticed regarding the amount of time spent outside the country for various purposes connected to their studies and career. Men tend to spend more time abroad than women, and the difference has become even more pronounced in 2005. In the case of men, the mean difference increased from 0.16 in 2000 to 0.47 in 2005. Similarly, there seems to be a downward trend in the number of women taking their degree from a developed country. The average for women obtaining their degree from a developed country decreased from 0.03 in 2000 to 0.01 in 2005, while for men it increased from 0.07 to 0.09. The age of these Malayali scientists working in both academic and government research sectors is generally between 40 and 50, but there are also many in retirement age group of 50-60 for both men and women. More men seem to have entered the science profession in 2005 as the youngest age group of 20-30 showed more male presence. All indicators except the highest degree point to a dominance of men in the research sectors of Kerala.

ICT Access and Use

Ynalvez et al (2005) observed that personal computer technology predates Internet technology and is now considered its precondition, though mobile phone access to the Internet is changing this. By 2005 almost all respondents had access to a computer at the workplace (96.9%), as against 85.8 per cent in 2000. Further, the percentage of scientists who have access to computers in their office doubled. In 2005 only half of all individuals have a computer in their personal office, but the number of persons sharing a computer has come down substantially from 5.38 in 2000 to 2.79 in 2005. Computer sharing remained more or less the same at home (3.06 and 3.24 in 2000 and 2005 respectively).

Table 1

Personal computer access and use profile of respondents

	Year	*Total*		*Sex*		*Sector*	
		No.	*%*	*M*	*F*	*Academic*	*Government*
At Work							
1. Have a PC at Work	2000	260	85.8	65.3	31.5	41.5	44.9
	2005	255	96.9	61.7	35.2	51.3	45.6
2. If yes, the year PC was first available	2000	260	1995	1994	1995	1996	1994
	2005	255	1997	1997	1998	1997	1998
3. If yes, is the PC located in the office	2000	260	25.4	21.2	4.2	6.6	18.8
	2005	254	51.6	42.2	9.4	31.5	19.7
4. If yes, number of people using PC Including R	2000	258	9.03	7.74	11.83	12.72	5.71
	2005	255	7.01	5.12	10.33	6.91	7.04
5. If yes, is the PC connected to the Internet	2000	118	45.6	34.4	11.2	11.6	34
	2005	232	91.7	59.3	32.4	48.6	43.1
At Home							
6. Have a PC at Home	2000	155	51.2	32	19.2	28.7	22.5
	2005	238	92.6	56.4	36.2	47.5	45.1
7. If yes year when PC was first acquired	2000	155	1998	1998	1998	1998	1997
	2005	238	2000	2000	2000	2000	2000
8. If yes no. of people using PC including R	2000	155	3.11	3.01	3.24	3.23	2.91
	2005	238	3.23	3.24	3.42	3.13	3.34

contd...

contd...

9.	If yes, is home PC connected to the net	2000	155	74.2	48.4	25.8	41.3	32.9
		2005	195	81.3	49.5	31.8	41.3	40
PC Use								
10.	No. of hours R uses PC in a week(at home/office)*	2000	132	0.96	1.24	1.12	1.23	1.43
		2005	242	2.97	3.12	2.54	3.11	2.73
11.	How often R uses PC for fun/play	2000	296	2.41	2.73	2.02	2.43	2.92
		2005	259	3.52	3.43	3.51	3.42	3.53
12.	How comfortable does R feel using PC	2000	296	3.41	3.43	3.51	3.51	3.32
		2005	261	1.32	1.23	1.42	1.23	1.03

* 0=Not at all, 1= Less than 1 hr, 2= between 1 hr and 5 hrs, 4=between 10 and 20 hrs, 5=over 20 hrs.

A gender analysis reveals a slight increase in the percentage of women having a computer in the office (31.5% in 2000 to 35.2% in 2005). The percentage of respondents who possessed a personal computer at home increased substantially from 51.2 per cent in 2000 to 92.6 per cent in 2005. However, the gender gap has increased from 12 per cent to 20 per cent in favour of men. Table 1 also shows an increase in the use of computers, as distinct from access. Line 10 shows that academic scientists now spend more hours each week using computers than government scientists, whereas in 2000 lower levels of computer use were reported by university scientists in Kerala (Ynalvez et al. 2005).

Line 5 of Table 1 reveals the great leap in Internet connectivity in the workplace. There was a doubling of Internet-connected computers in the workplace in the five years from 2000 to 2005. A corresponding growth in the home computers was not seen, although in 2000 a substantially higher percentage of home computers reported Internet connectivity (line 9 of Table 1). With a definite policy of adopting ICT as an integral part of its developmental initiative in Kerala, institutions in general were likely to obtain organizational Internet connectivity by 2005. The Information Technology Industry Policy Document by the Department of Information Technology, Government of Kerala in December 2001 stated in its preamble that the blueprint for IT development has been formulated in the context of emerging developmental trends and imperatives that are relevant to the growth strategy of the State. Notwithstanding the huge improvement, compared to their male counterparts, female scientists in Kerala still lag in having access to

Internet connectivity both at the work place and home. However, the primary focus of our analysis is not the use of computers as stand-alone tools, but rather their potential to facilitate global science. Hence our next question concerns the extent to which scientists make use of email and web browsing.

Table 2

Email Access and Use Profile of Respondents Across Sex and Sector

	Year	*Total*		*Sex*		*Sector*	
		No.	*%*	*M*	*F*	*Academic*	*Research*
E mail Access							
1. Whether R has ready access to email	2000	261	86.1	59	27.1	41.6	44.5
	2005	238	92.6	56.4	36.2	47.5	45.1
2. Whether R is currently using email	2000	253	83.5	57.6	25.9	38.9	44.8
	2005	245	96	62.4	33.6	52.2	43.8
3. Last time R sent email	2000	251	1.82	1.62	2.23	2.12	1.64
	2005	257	1.53	1.31	1.82	1.43	1.62
4. Year R first used email	2000	251	1997	1997	1997	1998	1997
	2005	254	1999	1998	2000	1999	2000
5. R has been unable to access email for at least one week	2000	87	34.8	26.4	8.4	12.0	22.8
	2005	94	37.0	22.4	14.6	20.1	16.9
6. Primary reason for not able to access email	2000	86	1.31	1.24	1.52	1.54	1.22
	2005	144	1.24	1.12	1.33	1.33	1.13
E mail Use							
7. If R wanted to send email R most likely do?	2000	86	1.30	1.24	1.50	1.48	1.21
	2005	257	1.82	1.85	1.76	1.79	1.86
8. No.of email sent in a typical week	2000	252	1.81	1.81	1.79	1.77	1.84
	2005	251	2.51	2.78	2.04	2.66	2.54
9. How many of these are research related	2000	250	1.81	1.94	1.53	1.49	2.09
	2005	250	2.26	2.50	1.84	2.41	2.09
10. Emails received in a week	2000	248	1.94	2.08	1.61	1.65	2.18
	2005	263	2.63	2.86	2.23	2.81	2.43
11. Hours spend for email every week	2000	247	1.49	1.58	1.29	1.35	1.62
	2005	248	1.92	2.01	1.76	1.49	2.00

contd...

contd...

Email Practices							
12. Discussed S&T issues	2000	78	30.83	24.11	6.72	13.04	17.78
	2005	115	44.92	33.50	11.32	21.87	23.04
13. Discussed research with someone in Kerala	2000	77	30.43	24.11	6.32	16.20	14.22
	2005	175	68.09	42.80	25.29	34.24	33.85
14. Discussed research with someone outside Kerala, within India	2000	73	69.52	50.47	19.04	20.00	49.52
	2005	194	75.49	53.31	28.18	41.63	33.85
15. Discussed research with someone in developed world	2000	146	57.70	44.66	13.04	21.34	36.36
	2005	139	54.29	40.23	15.06	30.07	24.21
16. Started professional relationship with someone met on Net	2000	88	34.92	27.77	7.14	14.68	20.23
	2005	79	30.73	23.73	7.00	15.95	14.78
17. Continued correspondence in Net with someone met personally	2000	180	71.42	53.96	17.46	29.36	42.06
	2005	178	69.53	47.65	21.87	38.67	30.85
18. Discussed proposal with funding agency	2000	128	50.59	37.54	13.04	20.55	30.03
	2005	154	59.92	40.85	19.06	32.68	27.23
19. Submitted and reviewed manuscript for journal	2000	65	25.69	20.16	5.53	7.11	18.58
	2005	158	61.48	42.02	19.07	35.41	26.07
20. Email Use and diversity index	2000	303	4.18	4.59	3.4	3.8	4.7
	2005	261	5.19	5.8	4.3	5.3	5.1

*0=Not at all, 1= Less than 1 hour, 2= Between 1 hour and 5 hours, 4=Between 10 and 20 hours

Table 3

Web Access and Use Profile of Respondents Across Sex and Sector Combination for Two Phase

		Year	*Total*		*Sex*		*Sector*	
			No.	*%*	*M*	*F*	*Academic*	*Research*
1.	R has ever used web browser	2000	199	65.68%	45.54%	20.13%	29.70%	35.97%
		2005	238	92.25	59.69	31.78	50.00	42.25
2.	Last time R browsed the web	2000	199	2.03	1.80	2.50	2.44	1.69
		2005	240	1.40	1.25	1.70	1.38	1.45

contd...

contd...

3.	Year R first used web	2000	199	1996	1998	1998	1996	1998
		2005	239	1999	1999	2000	1999	2000
4.	Hrs.R use web in a typical week	2000	197	1.63	1.69	1.48	1.44	1.78
		2005	238	2.20	2.35	1.91	2.30	2.08
FREQUENCY OF WEB ACCESS (1=daily, 2=weekly, 3= monthly, 4=less than once monthly, 5=never)								
5.	From home	2000	199	3.68	3.64	3.77	3.48	3.85
		2005	226	2.35	2.41	2.27	2.40	2.30
6.	From work	2000	199	2.70	2.34	3.51	3.30	2.20
		2005	239	1.64	1.46	1.99	1.59	1.70
7.	From public terminal	2000	198	4.73	4.75	4.57	4.56	4.78
		2005	237	4.50	4.49	4.51	4.63	4.34
8.	From Cybercafe	2000	199	4.90	4.87	4.98	4.99	4.83
		2005	237	4.50	4.49	4.51	4.63	4.34
9.	From Friend	2000	199	4.87	4.86	4.92	4.84	4.90
		2005	235	4.79	4.72	4.91	4.74	4.85
WEB USE(1=yes, -= no)								
10.	Ordered product/service for research	2000	82	41.21	29.96	11.56	14.07	27.14
		2005	124	51.24	35.95	15.29	29.75	21.49
11.	Created a web page	2000	48	24.12	18.09	6.00	9.55	14.57
		2005	54	22.41	16.18	6.22	12.45	9.96
12.	Conducted information search	2000	186	93.47	64.32	29.15	41.21	52.26
		2005	224	93.33	59.58	32.98	50.83	42.50
13.	Used E-journal	2000	107	53.77	41.21	12.56	21.11	32.66
		2005	166	69.17	50.42	18.75	42.92	26.25
14.	Used Data	2000	177	86.94	64.32	24.62	38.69	50.25
		2005	225	93.75	60.83	32.08	51.57	42.08
15.	Collaborated on a project	2000	70	35.18	29.15	6.03	14.07	21.11
		2005	122	50.83	35.83	15.00	30.42	20.41
16.	Found reference material	2000	180	90.45	62.81	27.64	41.71	48.73
		2005	233	97.08	61.67	34.58	52.08	45
17.	Accessed research reports	2000	162	81.41	58.29	23.12	59.30	46.73
		2005	212	88.33	58.33	29.17	77.92	39.17

contd...

contd...

18. Participated in online chat groups	2000	37	18.59	14.57	04.02	07.03	11.56
	2005	45	18.67	14.11	04.15	09.54	09.13
19. Used online job listings	2000	19	09.55	06.03	03.52	05.02	04.52
	2005	41	71001	14.11	02.90	09.96	07.05
20. Used online maps	2000	41	20.60	17.59	03.02	07.03	13.57
	2005	112	46.47	35.27	11.20	27.80	18.67
21. Downloaded software	2000	67	33.67	27.14	06.53	10.56	23.12
	2005	130	53.94	41.08	12.86	30.29	23.65
22. Published papers	2000	23	11.56	10.55	01.00	04.02	07.54
	2005	109	45.23	31.95	13.28	26.97	18.20
23. Web use and Diversity Index	2000	303	6.03	6.4	5.17	5.51	6.47
	2005	261	7.46	8.03	6.55	7.86	7.02

*0=Not at all, 1= Less than 1 hour, 2= Between 1 hour and 5 hours, 4=Between 10 and 20 hours.

Gender and sectoral differences in Internet access and use are our main focus in this analysis. Table 2 shows several dimensions of email use in both the study periods. We present percentages for current email use and ready access, as well as averages for email and web diversity and experience. The first line in Table 4 shows that email access for the scientists in Kerala, though pervasive, is not yet complete even in 2005. Only 92.6 per cent reported ready access to email, a 6 per cent increase since 2000. Sectoral differences are extremely small, but the digital divide in terms of gender is still evident: in the case of email access the gender gap has narrowed but remains at 20 per cent. Compared to 2000, small differences are apparent in the extent to which scientists define themselves as current email users across most social and professional dimensions in the later phase. Indeed, in 2005 there were virtually no differences among social categories except by gender. The reason is likely to be the sheer pervasiveness of email use; over 90 per cent of every category of respondent defined themselves as current email users.

Line 20 of Table 2 shows that five out of the nine possible email activities were carried out by the respondents in 2005. Compared to the survey in 2000 there is a general increase in the diversity of use of this technology, but for women respondents this average is slightly fewer (4.3 only). Although this increase is not large, for a period of five years, disparities in email use diversity still favour men. As evident from the fifth column of line 23 of Table 3, a similar pattern of disparity is observed for diversity in the use of the World Wide Web. Disparities in both email and web use experience also tend to follow the same lines as Internet practice. Those with greater experience and practice tend to be men with doctorates or degrees from developed countries, who are involved in professional organizations and have PCs located in their personal offices. Given these general results, we observe that the distribution and variation in Internet practice and experience, together with measures of ready access and current use, provide more reliable indicators of the current digital inequality in Kerala.

Table 4

Change in Publication Productivity by Gender and Sector

	Year	*Total*	*Mean*	*Gender*		*Sector*	
				Male	*Female*	*Academic*	*Research*
Papers in foreign journals	2000	303	1.10	1.08	1.12	1.44	0.72
	2005	261	3.84	5.52	1.07	5.66	1.82
Articles in national journals	2000	303	2.21	2.98	0.92	1.08	3.45
	2005	261	4.84	5.15	4.39	4.12	5.64

The rate of local and foreign publication over time is an important facet in examining the globalizing effects of the Internet. Publication productivity shows a general increase between 2000 and 2005 but with important differences by gender. While men increased both local and international productivity (as measured by the number of articles during the past five years in domestic or foreign journals), women exhibit an increase in local productivity but a decrease in international productivity. While we caution that these results are preliminary, they are cause for concern.

Discussion

We first examined change in the degree to which scientists in the developing world have adopted the information and communications technologies that are now a prerequisite for participation in global scientific communities. We addressed the diffusion of both hardware (computers) and connectivity (the Internet). Apart from variations by gender and organizational setting (sector), our longitudinal analysis highlighted a dramatic increase in the use, access, practice and experience of ICT (adoption). In 2000 'personalized' computers were not translated into a reality, although they were a pre-requisite for the use of the Internet. While most scientists then had access to a computer at work, only a minority could boast of a PC in their personal office. PCs were not so much 'personal computers' as 'public computers: over nine people had to use each computer. Most described themselves as Internet users, but few had ready access and most had little practice or experience.

By 2005 the situation had changed substantially in the use, access, practice and experience of ICT. Computers with connectivity were available to nearly all at both home and office. Hence the substantial discrepancy observed in 2000 between scientists' view of themselves as Internet users and the extent to which they actually had the opportunity to use the Internet seems to have declined significantly. The a self-defined email user in Kerala is now capable of engaging in a scientific collaboration that requires reliable or continuous access. Earlier sharing of resources, especially with a large number of other scientists, translated into the low individual usage that indirectly affected the development of sophisticated Internet practice and relevant Internet experience. This is now likely to have been overcome to a great extent.

As regards productivity, in terms of publications in regional and foreign journals, our 2000 study revealed academics were less productive than government research scientists. Ynalvez et al (2005) concluded that disparities in ready access and current use did not generally influence either foreign or local productivity. More importantly, they found the digital divide that had a direct bearing on scientific output was the divide that pertained to practice and experience, not access and use. By 2005, although the digital divide in terms of Internet practice and experience seem to have disappeared. We had hoped that the Internet would reduce the gender divide, for instance, but preliminary results do not seem to support this idea. While there are indications of a general increase in productivity, but it is evident primarily for men. From 2000 to 2005, differences in international productivity by gender became sharper, as there seems to be an actual reduction in foreign publication by female scientists.

These preliminary results suggest gender remains a durable source of inequality in Internet adoption and productivity. Among the few non-users, most were women. Men continued to spend more time in research and the mean difference increased from 11.76 to 12.75). But education seems to have taken a dramatic turn, with slightly higher percentages of women having doctorate. Further there was an increase in the number of women professionals in universities. Nevertheless educational and organizational localism were likely to be more pronounced: women were less likely to acquire an advanced degree from abroad. Other indicators (not shown here) also point to the increasing degree of 'localism' in the organizational, professional and productivity dimensions in the lives of female scientists, even as they communicate with the larger world via the Internet. Their local productivity has increased, but their international productivity has decreased. The results are consistent with Miller et al (2006), who suggest the shift to Internet-based science has not yet translated into overall occupational benefits for women. Although the Internet has enabled women scientists to circumvent isolation, it has increased organizational localism.

Women were more likely to believe that the Internet did make a difference in their professional lives, though women scientists in the academic sector did not subscribe to this proposition. For both men and women the importance of web and mail seem to have increased substantially, but men were more likely to see the importance of Internet in research. Ynalvez et al (2005) noted increasing disparities among social characteristics as one moved from current use, to ready access, then to practice and finally to experience. They concluded that the digital divide pertained more to aspects of practice and experience than adoption in any simple sense. Our 2005 results reinforce their contention that current use and access may become less relevant aspects of the digital divide in the future. As Kerala develops more reliance on new information and communication technologies, the concept of the digital divide moves away from simple inequalities in hardware and software access/ use and toward the more complex inequalities of user interaction skills.

Our results revealed that the efficacy of Internet as a research tool is likely to gain momentum as the issues of space and identity are addressed, with the greater temporal and spatial autonomy that come from workplace privacy. Most men and women considered that matters have improved owing to the Internet and see greater prospects in the future. Women were more likely to believe that Internet did make a difference. So there is greater chance of transforming Internet connectivity into a collaborative research tool, making science a truly global enterprise. As the concept of 'reagency' suggests, the Internet has emerged as an essential technology in generating, producing and disseminating scientific knowledge in Kerala.[3]

Notes

1. Specifically, for instance: Has been a member of science and technology (S&T) discussion group, has sent a message to an S&T discussion group, discussed research with someone in a developed country, started a professional relationship with someone met on the Internet, discussed research proposal with funding agencies, submitted or reviewed manuscripts for journals.
2. The specific questions include: order a product or service for research, created a web page, conducted an information search, used an electronic journal, acquired or used data, collaborated on a scientific project, found and examined reference materials, accessed research reports or scientific papers, participated in chat groups, used job listings, used maps, downloaded software, published a paper.
3. This material is based upon work supported by the National Science Foundation under Grant No. 0113545.

References

Allen, R.S. 1991. "Physics information and scientific communication: information sources and communication patterns". *Science and Technology Libraries*. 11(3) 27-38

Allen, T.J. 1977. *Managing the Flow of Technology: Technology transfer and the Dissemination of Technological Information within the R & D Organization*. Cambridge, M.A: MIT Press.

Barjak, F. 2004a. *On the Integration of the Internet into Informal Science Communication*. (Series A: Discussion paper DPW 2004-02) Olten: Switzerland: University of Applied Sciences Solothurn Northwest Switzerland. Retrieved April 23, 2007 from http: //www. fhso.ch/pdf/publikationen/dp04-02.pdf

Barjak, F. 2005. *Research Productivity in the Internet Era.* (Series A: Discussion paper DPW 2005-01) Olten: Switzerland: University of Applied Sciences Solothurn Northwest Switzerland. Retrieved April 23, 2007 from http: //www. fhso.ch/pdf/publikationen/dp05-01.pdf

Crane D. 1972. *Invisible Colleges: Diffusion of Knowledge in Scientific Communities.* Chicago: University of Chicago Press.

Davidson, T., R. Sooryamoorthy, and W. Shrum 2002. "Kerala Connections: Will the Internet Affect Science in Developing Areas? in B. Wellman and C. Haythornthwaite *The Internet in Everyday Life*. USA: Blackwell.

Dedijer, Steven. 1963. "Underdeveloped Science in Underdeveloped Countries", *Minerva.* 2(1), 61-81

Di Maggio, P., E. Hargittai, W.R. Neuman, and J.P. Robinson. 2001. "Social Implications of the Internet". *Annual Reviews of Sociology.* 307-336.

Finholt, T.A. and Olson, G.M. 1997. From laboratories to collaboratories: A new organizational form for scientific collaboration (Electronic version) *Psychological Science* 8(1), 28-36

Fry, J. 2006. "Scholarly research and information practices: A domain analytic approach". (Electronic version). Information Processing and Management. 42 (1), 299-316.

Franke, R.W. and B. H. Chasin. 1991. *Kerala: Radical Reform as Development in an Indian State.* San Francisco: Institute for Food and Development Policy.

—. (2000). *Is the Kerala Model Sustainable? Lessons from the Past, Prospects for the Future.* London: Zed Books.

Garvey, W.D. 1979. *Communication, the Essence of Science: Facilitating Information Exchange among Librarians, Scientists, Engineers, and Students.* New York: Pergamon Press

Garvey, W.D. and Griffith B.C. 1967. "Scientific Communication and Social System". (Electronic version). *Science.* 157 (3792), 1011-1016.

Glaser, J. 2003. "What internet use does and does not change in Scientific Communities." *Science Studies.* 16 (1), 38-51.

—. 1972. Communication and Information processing within scientific disciplines – Empirical findings for Psychology. (Electronic version). *Information Storage and Retrieval.* 8 (3), 123-136.

Gresham, John Jr. 1994. From invisible college to cyberspace college: Computer conferencing and the transformation to informal communication networks. (Electronic version). *Interpersonal Computing and Technology: An Electronic Journal for 21st Century.* 2 (4), 37-52.

Hesse, B.W., Sproull, L.S., Kiesler, S.B. and Walsh J.P. 1993. "Return to Science: Computer Networks in Oceanography" (Electronic version). *Communication of the AC. 36 (8), 90-101.*

Kaminer, N.B. and Braunstein, Y.M. 1998. Bibliometric analysis of the impact of internet use n scholarly productivity. (Electronic version). *Journal of the American Society for Information Science.* 49 (8), 720-730.

Kauzes, R.T. 2000. Electronic collaboration in environmental and physical sciences research. In S. H. Koslow and M.F. Huerta (Eds.) *Electronic Collaboration in Science.* Mahwah, NJ: Lawrence Erlbaum.

Kling, R., Mckim G. and King, A. 2003. "A bit more to it: Scholarly communication forums as socio-technical interaction networks". (Electronic Version). *Journal of the American Society for Information Science and Technology.* 54 (1), 47-67.

Lacy, W.B. and Busch L. 1983. Informal Scientific Communication in the Agricultural Sciences. (Electronic version) *Information and Processing Management.* 19: 4, 193-202.

Lievrouw, L.A. and Carley, K. 1990. "Changing patterns of communication among scientists in the era of telescience". *Technology in Society.* 12 (4), 457-477.

Lamb and Davidson. 2005. Information and communication technology challenges to professional scientific identity. (Electronic version). *Information Society,* 21 (1), 1-24.

Liebscher, P., Abel, E.G., and Denman D.W. 1997. Factors that influence the use electronic networks by science and engineering faculty at small institutions. Part II. Preliminary use indicators. (Electronic version). *Journal of the American Society for Information Science,* 48 (6), 496-507.

Mulla, K.R. and Chandrashekara, M. 2006. "Internet Users: Mysore University Campus (India)" *SRELS Journal of Information Management,* 43 (3) September, 243-263.

Parayil, Govindan. 1996. "The 'Kerala Model' of Development: Development and Sustainability in the Third World". *Third World Quarterly* 17 (5), 944-957.

Robinson, J. 2003. "Introduction to Issues 4 and 5 Digital Divides: Past, present and future". *IT & Society.* 1(5), i-xiv.

Russell, J.M. 2001. "Scientific communication at the beginning of the twenty-first century" (Electronic version). International Social Science Journal, 53 (168).

Salam, Abdus. 1996. "The isolation of the scientist in developing countries", *Minerva.* 4(4), 461-65.

Shrum, Wesley. 2005. "Reagency of the Internet, or, How I became a Guest for Science," Social *Studies of Science* 35(5), 723-754.

Sondergaard, T.F., Anderson J. and Hjorland B. 2003. Documentation and the communication of scholarly information – revising and updating the UNISIST model (Electronic version) *Journal of Documentation*. 59 (3), 278-320.

Uimonen, P. 2001. *Transnational.Dynamics@Development.net: Internet, Modernization and Globalization.* Stockhlom, Sweden: Elanders Gotab.

Walsh, J.P. and Bayman T. 1996a. "The virtual college: Computer mediated communication and scientific work" (Electronic version). *Information Society*. 12 (4), 343-363.

Walsh, J.P., Kucker, S., Maloney, N.G., and Gabby, S. 2000. "Connecting minds: Computer mediated communication and scientific work" (Electronic version). *Journal of the American Society for Information Society*. 51 (14) 1295-1305.

Walsh, J.P. and Maloney, N.G. 2002. "Computer network use, collaboration structures and productivity" in P. Hinds and S. Kielser (Eds.). *Distributed Work*. Cambridge MA: MIT Press.

Ynalvez Marcus, R. Duque, P. Mbatia, R. Sooryamoorthy, A. Palackal, and W. Shrum. 2005. "When do scientists "Adopt" the Internet? Dimensions of Connectivity of Developing Areas". *Scientometrics*. 63 (1).

11

Information Technology for Kerala's Development

JAYAN JOSE THOMAS

This article discusses the possibilities of making use of information technology (IT) for Kerala's development. Kerala has made remarkable achievements in social spheres, particularly in the fields of land reform, health, and universal school education (Ramachandran 1996). In 2001, the literacy rate among females above the age of seven in Kerala was 87.9 per cent, compared to 54.3 per cent in India as a whole. Despite these social achievements, income growth and industrialization in Kerala lagged behind the corresponding Indian averages (Thomas 2005a).

Given its large, educated workforce, Kerala possesses the potential to achieve faster economic growth in the emerging 'new economy' or 'knowledge economy' characterized, most importantly, by the extensive use of information and communication technologies. However, the record of Kerala's performance in the IT industry so far is not very encouraging. As Table 1 shows, the share of Kerala in total exports of IT software and services from India in 2003-04 was only 0.5 per cent (against Kerala's share in India's population of 3.1%). Karnataka, Tamil Nadu and Andhra Pradesh, the neighbouring states have shares of 33.4 per cent, 14.0 per cent and 9.7 per cent respectively, in the total IT software and services exports from India in 2003-04 (see Table 1). The major cities in Kerala including Kochi and Thiruvananthapuram have fared far below their potential in the IT software and services industry. Their record in the growth of IT sector is way behind that of other major south Indian cities, importantly Bangalore, Chennai and Hyderabad.

Table 1

Shares of Kerala and Other Selected Indian States in Total Export of IT Software and Services from India, 2003-04

States/Union Territories	*Percentage of Total Exports*
Karnataka	33.4
Maharashtra	17.2
Tamil Nadu	14.0
Andhra Pradesh	9.7
Haryana	8.6
Uttar Pradesh	6.1
Delhi	6.0
West Bengal	2.8
Kerala	0.5
Punjab	0.3
Gujarat	0.3

Source: Lok Sabha Unstarred Question No. 3136, dated 13-8-2004, accessed from <www.indiastat.com>

Kerala does possess some advantages in the IT software and services industries. Kochi is the only location in India for two international submarine cable links (SEA-ME-WE-3 and SAFE). Recently Videsh Sanchar Nigam Limited (VSNL) inaugurated the SAT-3/WASC/SAFE submarine cable station in Kochi, one among the 16 landing points across the world for this submarine cable. The cable system has a capacity of 120 Gbps (Giga bytes per second), which is powerful enough to transmit a total of 5.8 million telephone channels simultaneously.[1] Given all these, sourcing bandwidth from Kochi is cheaper than from most other locations in the country.[2] Technopark in Thiruvananthapuram was one of the first dedicated IT parks in the country when it was established in 1995. Operation costs including the cost of office space is lower in Kochi and Thiruvananthapuram compared to the corresponding costs in major metropolitan cities in India. There is no shortage of skilled professionals in Kerala. NASSCOM, a countrywide association of software companies, has rated Kochi as the second best location in India, after only Hyderabad, for bandwidth intensive, ICT-enabled services.[3] Why, despite such advantages, is Kerala lagging behind in the IT software and services industries? Knowledge-based industries are characterized by high degree of increasing returns. These industries exhibit a tendency towards geographical clustering around certain locations. With geographical clustering, there arises opportunities for exchange of tacit knowledge among related firms in the knowledge industry. There will also be opportunities for 'collective learning' among these firms (Maskell and Malnberg 1999).

Does Kerala offer an environment that is conducive for the clustering of knowledge-intensive industries? Industrial structure in Kerala is characterized by technological backwardness and relatively low diversification. Two industries – chemicals and rubber-based industries – dominate the factory sector of Kerala with respect to the value of output produced. Two other industries – cashew processing and *beedi* (local cigarette) making, both of which employ highly traditional technologies – dominate employment in the factory sector of Kerala. Chemicals-based industries were first established in Kerala in the 1930s, when Kerala enjoyed some advantages in the availability of hydroelectricity from the Pallivasal Hydroelectric Project. Despite the low level of local demand for chemical products and the absence of raw material required in the production of chemicals in Travancore, the Travancore administration of the 1930s and 1940s encouraged power-intensive, heavy chemical industries in the Alwaye-Kochi area, offering hydroelectricity at cheap rates. After 1950, direct investment in industry in Kerala by the Central Government has been relatively low. State Government in Kerala – whose degrees of freedom with respect to intervention in industrial development have been curtailed by the low level of fiscal transfers from the Centre to the States – continued to encourage investments in large-scale, chemicals-producing industries. However, Kerala's advantages in the availability of hydroelectricity were short-lived such that by the 1980s there was a chronic power-deficit. Power rationing and power cuts have been the most important constraint to the growth of large-scale industry in Kerala (Thomas 2003; Thomas 2005a; Thomas 2005b).

On the other hand, cities such as Bangalore and Hyderabad received large investments by the public and private sectors over the period of planned development in India. These investments have been in technologically advanced industries – in industries such as electrical and electronics, aircraft manufacturing and machine tools in the case of Bangalore. Several academic and research institutions of national importance have been set up in these cities. Engineering colleges and other technical institutes have also sprung up in these cities in large numbers. All these have created a favourable ecosystem for the growth of technologically advanced industries and have greatly facilitated the growth of IT and ITES industries in Bangalore and Hyderabad.

For cities in Kerala to catch up with Bangalore, Hyderabad or Chennai in the IT industry, a huge effort with respect to investment is required. The government, in collaboration with industry and academic institutes or universities (a 'triple helix' mode of interaction) must guide investments into Kerala in information technology and biotechnology. Thiruvananthapuram and Kochi as well as Kottayam and Kozhikode possess the potential to emerge as knowledge hubs. These cities are home to major universities in Kerala, and there is a large supply of skilled professionals in these cities. With respect to biomedical sciences, the large numbers of health professionals in Kerala including nurses and doctors is a significant advantage. What is required of the government is to set up – independently or in alliance with the private sector – major centres of research and learning in the new technologies. Research in biotechnology and biomedical sciences in Kerala must focus on the specific needs of the region. For example, carrying out research on processes to enhance value addition in spices, rubber and marine products will be an important step towards restoring health to the agrarian economy of Kerala. Into an environment conducive for high technology industries that is built by government investment, private investment will flow in. It is important to note that without such a major push through government intervention, Kerala will not be able to compete with other centres of agglomeration in high-technology industries in India.

A major advantage of Kerala is that economic activities using new technologies can be pursued extensively in rural areas of the State, as easily as in the bigger cities. Geographical configuration of

Kerala transcends standard rural-urban boundaries. There is high degree of rural-urban homogeneity in Kerala in terms of social achievements and economic opportunities. Kerala has a large number of small towns, which have been the centres for wider diffusion of economic activity in the State. Almost every village in Kerala is connected by motorable road. Among Indian states, Kerala has the highest number of telephones per 100 rural population. In rural areas of Kerala, in June 2005, there were 9.7 telephones per 100 population; the corresponding figures for the rural areas of India as a whole and rural areas of Punjab were, respectively, 1.7 telephones and 5.3 telephones per 100 population. Telephone density in urban Kerala, too, is relatively high; in June 2005, it was higher than the telephone density in the city of Mumbai (see Table 2).

Table 2

Telephones per 100 Population, Rural and Urban Areas, Selected States/Telecom Circles in India, June 2005

State	*Urban*	*Rural*	*Overall*
Kerala	47.6	9.7	19.5
Himachal Pradesh	78.1	6.8	14.2
Punjab	51.6	5.3	22.9
Haryana	29.2	2.9	11.1
Tamil Nadu - Chennai	23.1	2.9	11.5
Gujarat	30.1	2.6	13.2
Maharashtra - Mumbai	27.7	2.6	10.7
Karnataka	31.3	2.5	12.7
Andhra Pradesh	30.2	2.4	9.9
West Bengal - Kolkata	17.1	1.1	3.3
Uttar Pradesh	18.9	0.5	4.4
Kolkata	25.1	—	25.1
Chennai	48.0	—	48.0
Delhi	52.1	—	52.1
Mumbai	45.8	—	45.8
Total	28.3	1.7	9.4

Source: Lok Sabha Starred Question No. 244, dated 10.8.2005 cited in www.indiastat.com accessed on 10 March 2006.

A field-study based research conducted by this author in July-August 2004 showed that rural areas in Kerala are much better prepared for ICTs-aided development compared to rural areas in other Indian states (Thomas 2006). The study included sample surveys of households in selected villages in Kuppam in Andhra Pradesh and Malappuram in Kerala. The ongoing i-community project in Kuppam and Akshaya Project in Malappuram seek to provide access to ICTs to the local population through information centres, with the aim of enhancing developmental opportunities in agriculture, health and education. Among the surveyed households in Malappuram in Kerala, where literacy rate was 96.4 per cent, the Akshaya Project led to widespread use of computers, even by women and also by the socially backward class of scheduled tribes. Among the surveyed households in Kuppam in Andhra Pradesh, where literacy rate was 63.3 per cent, the i-community project produced far less impressive results with respect to the use of computers.

The Akshaya Project in Kerala was first launched in Malappuram district. Recently, the project has been rolled out in seven more districts in Kerala: Ernakulam, Thiruvananthapuram, Alappuzha, Idukki, Kannur and Wayanad. According to officials at Kerala IT mission, the Akshaya *e-kendras* have trained 600,000 persons in the use of computers by May 2003. These *kendras* also serve as centres for the delivery of government services and e-services. The Akshaya Project aims to reach the benefits of information technology to 6.4 million families in Kerala; it aims to familiarize at least one person from each family in Kerala with computers.

Wide diffusion of information technology in rural areas if complemented with the practice of democratic decentralization will be a powerful means for energizing the rural economy of Kerala. Activities ranging from agro-processing to tour operation can be carried out under *panchayats* with the effective use of IT. Along with this, the emergence of high-technology centres or 'knowledge' centres in major towns in Kerala will put the State's economy into a higher growth path.

Notes

1. <http: //dotcomtechno.com/whyindia.html>
2. On Kerala government's IT policy see <http://kerala.gov.in/annualprofile/it.htm>
3. See *Advantage Cochin Special Economic Zone* from The Development Commissioner, Cochin Special Economic Zone, Kochi (NASSCOM is National Association of Software and Service Companies).

References

Maskell, Peter and Malnberg, Anders. 1999. "Localised learning and industrial competitiveness". *Cambridge Journal of Economics,* 23 (2), 167-85.

Ramachandran, V. K. 1996. "On Kerala's Development Achievements". In Drèze, Jean and Sen, Amartya (eds.) *Indian Development: Selected Regional Perspectives.* New Delhi: Oxford University Press.

Thomas, Jayan Jose. 2003. "Labour and Industrialization in Kerala". *Indian Journal of Labour Economics.* 46 (4), 575-92.

—. 2005a. *State Policy, Industrial Structure and Industrialization: The Case of Kerala,* Ph.D. thesis, Indira Gandhi Institute of Development Research, Mumbai. (Mimeo).

—. 2005b. "Kerala's Industrial Backwardness: A Case of Path Dependence in Industrialization". *World Development,* 33 (5), 763-83.

—. 2006 "Informational Development in Rural Areas: Some Evidence from Andhra Pradesh and Kerala". In Parayil, Govindan (ed.) *Political Economy and Information Capitalism in India: Digital Divide, Development and Equity.* Basingstoke, UK: Palgrave Macmillan, 109-132.

Index

Aart, Scholte, J., 152, 157

Abbate, Janet, 178

Abraham, Margaret, 178

Acosta-Belen, Edna and Christine E. Bose, 173, 175

Akshaya (Kerala's pioneering ICT dissemination initiative), 48-50, 52-53, 112, 123-27, 141-43
  activities of, 60-62
  features of, 54-60
  issues addressed by, 53-54

Alexander, W.M., 159

Allen, R.S., 205

Allen, T.J., 205

Amancio, L., 155

Antin, J.D., 123

Arunachalam, S., 133

Arun, S. and T. Arun, 157, 159

Barjak, F., 202, 206, 207

Barry, Wellman, 22

Basu, A. and B.S.V. Kumar, 153, 158

Basu, A. and R. Aggarwal, 153, 166

Bhatnagar, S., 110

Bihar
  computerization initiatives in, 113-14

Boserup, Esther, 175

Campion, P. and W. Shrum, 156, 174, 178, 179, 180, 183, 186, 187, 188, 189

Chakravarthy, Radha, 155, 156, 179

Chanana, Karuna, 179

Chasin, B.H., 26

Cole, Jonathan R. and Harriet Zuckerman, 178
collaboratories, 207
computerization projects
  in Bihar, 113-14
  in Kerala, 115-16
Connell, R.W. and J. Wood, 155
Crane, D., 206, 207

Davidson, Theresa, R. Sooryamoorthy and W. Shrum, 174, 202
De Roy, O.C., 156
Dedijer, Steven, 201
Derne, Steve, 179
development
  concept of, 135-38
DiMaggio, P. et al., 202
Drori, G., Meyer, F. Ramirez and E. Schofer, 153, 154
Drucker, Peter, 174
Duncombe and Heeks, 137
Duque, R.B. et al., 153, 157

Eapen, Mridul and Parveena Kodoth, 182
e-governance, 109
  in India, advent of, 110-112
Ekdahl, P. and L. Trojer, 157
Escobar, Arturo and Wendy Harcourt, 177
Escobar, Arturo, 135, 136, 157, 168, 177

Ferguson, James, 135
Finholt, T.A. and, G.M. Olson, 207
Flinders, M.V. and M.J. Smith, 128
Fox, M.F. and Scott J. Long, 178
Fox, M.F., 155, 178
Frame, J.D. and M.P. Carpenter, 153
Franke, R.W. and B.H. Chasin, 158, 181, 200
Franke, R.W., 26
Free Software education
  in Kerala, 104-5
Free Software, 77
  concept of, 77
  freedoms provides by, 77-78
  in business, 90-91
  in India, scenario of, 77
  in Kerala 78-94
FRIENDS (an ICT-enabled access point to government for improving service levels in Kerala), 43-45, 112, 116-21
  activities of, 119
Fry, J., 204

Gaillard, J., V.V. Krishna and R. Waast, 158
Gajjala, Radhika and A. Manidipudi, 173, 174, 176
Gajjala, Radhika, 173, 176
Garvey, W.D., 204
Garvey, W.D., and B.C. Griffith, 204, 206
Garvey, W.D. and Gottfredson, 206
George, K.K., 26, 181

Gerester, R. and S. Zimmerman, 137
Giddens, A., 157
Glaser, J., 206
globalization, 152
Gresham, John, Jr., 206
Gupta, Nomrate and A.K. Sharma, 155, 179, 189

Harcourt, Wendy, 177
Harding, Sandra, 177
Hart, Gillan, 137
Held, D., A.G. McGraw, D. Goldblatt and J. Perraton, 153, 170
Hesse, B.W., Sproull, L.S., Kiesler, S.B. and J.P. Walsh, 206, 207
Huyer, S. and M. Carr, 156, 157, 171
ICT dissemination projects
  access to, 156
  in Kerala, 43-50
  in rural India, 42
ICT-enabled information system
  in Kerala, 64; barriers in, 71; impact on productivity, 73-75
ICTs (Information and Communication Technologies), 41
  access and use, 210
  in India, 41-43
  in Kerala, 43
ICTUD, 137-38
Information Kerala Mission (IKM), 45-48
Information Society, 109, 133
information villages
  in Pondicherry, 42
Internet Age, 206
Internet and Science, 202
Internet for action
  concept of, 75
Issac, Thomas, 26
IT education
  in Kerala, 106
IT@School Project
  in Kerala, 99-101
Iyer, Ramanathaiyer, 26

Kaminer, N.B. and Y.M. Braunstein, 207
Kannan, K.P., 26
Katakam, A., 111
Kauzes, R.T., 201
Kelkar, Govind and Dev Nathan, 174, 176
Keller, Evelyn, Fox, 178
Kemelgor, C. and H. Etzkowitz, 155
Keniston, K., 133
Kerala Model of Development, 181
Kerala, 77
  Free Software in, 77
  ICTs in, 43
Khan, Sarbuland, 23
Khan, Waheed, Abdul, 21, 25
Kleiwachter, Wolfgang, 21
Kling, R., Mckim, G. and A. King, 204
knowledge economy, 223
Kumar, N. 155, 156
Kuriyan, R., 143, 146

Kyvik, Sven, and I. Larsen, 153, 166
Kyvik, Sven, and M. Teigen, 155, 178

Lacy, W.B. and L. Busch, 206
Lal, D., 135
Lamb and Davidson, 201
Lee, S. and B. Bozeman, 153
Liebscher, P., Abel, E.G. and D.W. Denman, 201
Lievrouw, L.A. and K. Carley, 205, 207
Lin, Nan, 154, 180
Long, J., 155
Luukkonen, T.O. Persson and G. Sivertsen, 157

Madon, S. and G.R. Kiran, 118, 122
Madon, S., 110, 123
Mann, C., 135
Maskell, Peter and Malnberg, Anders, 225
McElrath, Karen, 178
McMichael, P., 157
Mies, Maria, 175
Millennium Development Goals (MDGs)
of the UN, 135, 137
Miller, B.P. et al., 218
Mohan, R., 26,
Mohanty, Chandra Talpade, 174, 177
Morgall, Janine Marie, 173, 175, 176
Mukhopadhyay, Carol C. and Susan Seymour, 179
Mulla, K.R., and M. Chandrashekara, 202

Nederveen, Pieterse J., 153, 157
Nie, N.H., 23
Nie, N.H. et al., 23

Oommen, M.A., 26, 181

Pal, J. and G.R. Kiran, 123
Pal, J. et al., 143
Pal, Joyojeet, 174, 176
Palackal, A., M. Anderson, B.P. Miller and W. Shrum, 156, 157, 166, 189
Palackal, A. et al., 25, 28
paperless society, 70
Parayil, Govindan, 182, 200
Park, Kyunge Ae, 175
Patel, Reena and Mary Jane C. Parmentier, 176
patrifocality, 179
Portes, A., 154
Prahalad, C.K. and Allen Hammond, 140
Prahalad, C.K., 133
Putnam, R.D., 37

Quan-Haase, Anabel, 22

Ramchandran, V.K., 26, 181, 223
Ramu, G.N., 179
Renjini, D., 182
Ranson, Gillian, 178

Richards, Lynn, 182
Riessman, Catherine Kohler, 179
Robinson, J.M., 201
Rout, Usha Rani, Sue Lewis and Carolyn Kegan, 179
RRLT (CSIR's Regional Research Laboratory, Trivandrum), 64
  ICT-enabled information system in, 64-71
Russell, J.M., 202

Sachs, W., 135, 136
Salam, Abdus, 201
Schott, P., 152, 154
science communication, 204
Sen, Amartya, 26, 135
Shrum, Wesley et al., 25
Shrum, Wesley and S.R. Iyer, 181
Shrum, Wesley, 25, 26, 28, 156, 174, 202
Shukla, Archana, 179
Smith-Lovin, L. and J. McPherson, 154, 155
social capital, 154
Sondergaard, T.F., Anderson J. and B. Hjorland, 204, 205
Sooryamoorthy, R. and Wesley Shrum, 182
Sreekumar, T.T., 109, 110, 112
Stauffacher, Daniel, 21
Subrahmanyan, Lalita, 178, 179
Theil, S. van, 128
Thomas, Jayan, Jose, 223, 225
Toyama, K. et al., 145

Uimonen, P., 203

ViCTERS (the edusat initiative) in Kerala, 101-2
Vittal N., 120

Wajcman, Judy, 173, 176
Walsh, J.P. and N.G. Maloney, 207
Walsh, J.P., Kucker, S., Maloney, N.G. and S. Gabby, 207
Walsh, J.P. and Bayman, T., 207
women and technology, 175
Woolgar, S., 109
World Bank, 137
World Summit on the Information Society, 63

Xie, Y. and K.A. Shauman, 155, 178

Ynalvez, Marcus et al., 200, 211, 217, 218
Youngs, Gillian, 174